GUIDELINES FOR

Writing Effective Operating and Maintenance Procedures

GUIDELINES FOR

Writing Effective Operating and Maintenance Procedures

CENTER FOR CHEMICAL PROCESS SAFETY
AMERICAN INSTITUTE OF CHEMICAL ENGINEERS
345 East 47th Street, New York, New York 10017

345 East 47th Street
New York, New York 10017

Library of Congress Cataloging-in Publication Data
Guidelines for writing effective operating and maintenance procedures.
p. cm.
Includes bibliographical references (p. –) and index.
ISBN 0–8169–0658–0
1. Chemical engineering—Safety measures—Handbooks, manuals, etc.
2. Technical writing—Handbooks, manuals, etc. I. American Institute of Chemical Engineers. Center for Chemical Process Safety.
TP149.G845 1996
660'.2804'0684—dc20 96–33781
CIP

It is sincerely hoped that the information presented in this document will lead to an even more impressive safety record for the entire industry; however, the American Institute of Chemical Engineers, its consultants, CCPS Subcommittee members, their employers, their employers' officers and directors, and NUS Training Corporation and its employees, officers and directors disclaim making or giving any warranties or representations, including with respect to fitness, intended purpose, use or merchantability and/or correctness or accuracy of the information presented in this document. As between (1) American Institute of Chemical Engineers, its consultants, CCPS Subcommittee members, their employers, their employers' officers and directors, and NUS Training Corporation and its employees, officers and directors , and (2) the user of this document, the user accepts any legal liability or responsibility whatsoever for the consequences of its use or misuse.

CONTENTS

PREFACE

For over 30 years the American Institute of Chemical Engineers (AIChE) has been involved with process safety and loss control issues in the chemical, petrochemical, hydrocarbon process and related industries and facilities. AIChE publications and symposia are information resources for the chemical engineering and other professions on the causes of process incidents and the means of preventing their occurrences and mitigating their consequences.

The Center for Chemical Process Safety (CCPS), a Directorate of the AIChE, was established in 1985 to develop and disseminate technical information for use in the prevention of major chemical process incidents. With the support and direction of the CCPS Advisory and Managing Boards, a multifaceted program was established to address the need for Process Safety Management systems to reduce potential exposures to the public, facilities, personnel, and the environment. This program involves the development and publication of Guidelines relating to specific areas of Process Safety Management; organizing, convening and conducting seminars, symposia, training programs, and meetings on process safety-related matters, and cooperation with other organizations, both internationally and domestically, to promote process safety. The CCPS's activities are supported by funding and professional expertise by over 90 corporations. Several Government agencies also participate in CCPS's endeavors.

In 1989, CCPS published the *Guidelines for Technical Management of Chemical Process Safety*, which presented a model for Process Safety Management characterized by twelve distinct, essential and interrelated elements. The Foreword to that volume stated:

> For the first time, all the essential elements and components of a model of a technical management program in chemical process safety have been

assembled in one document. We believe the Guidelines provide the umbrella under which all other CCPS Technical Guidelines will be promulgated.

One of the elements of chemical process safety developed in the *Guidelines for Technical Management of Chemical Process Safety* is training and performance. Operating and maintenance procedures are essential to achieving proper training and safe, efficient performance of operating and maintenance tasks. This book was written to assist all persons involved with operating and maintaining chemical process facilities in establishing a system to write and implement procedures. It is not intended to serve as a compliance manual to meet the specific requirements of any mandates, rules, laws or regulations. It emphasizes those principles needed to write, implement, and control effective operating and maintenance procedures. This Guideline, while dealing primarily with the preparation and control of procedures, also treats some of the other issues of the Training and Performance element of chemical process safety. The purpose of the examples within this guideline is solely to demonstrate a technique or methodology. The examples are not complete procedures and therefore are not intended to be evaluated for technical accuracy or soundness. Other Guidelines dealing specifically with training and the actual operation and maintenance of chemical facilities are also currently being prepared.

ACKNOWLEDGMENTS

The American Institute of Chemical Engineers and the Center for Chemical Process Safety (CCPS) expresses its gratitude to all of the members of the Subcommittee on Writing Effective Operating and Maintenance Procedures for their unstinting efforts and technical contributions in the preparation of this Guideline. CCPS also expresses its appreciation to members of the Technical Steering Committee for their valuable advice and support.

The Chair of the Subcommittee on Writing Effective Operating and Maintenance Procedures was John A. McIntosh, III of Procter and Gamble. The Subcommittee members were Mr. Neil Maxson, Bayer Corporation (formerly Miles, Inc.); Lisa Loden and Tommy Northcutt, Lockheed Martin (formerly Martin-Marietta Energy Systems); Robert Kambach, BASF; William Fleming, Carmel Corona, Jr., and Vijay Surapaneni, Bristol-Myers Squibb Company; Ken Turnbull, Kathy Hartkey, Gus Wilson and Roman Wally of Texaco, Inc.; Peter Hughes of Novacor Chemicals Company, Ltd. and John R. Lockwood of BP America. Dr. Martin Gluckstein, P.E. was the CCPS staff liaison and was responsible for the overall administration and coordination of the project.

NUS Training Corporation, a Williams Knowledge Systems Company, Gaithersburg, MD, was the contractor responsible for preparing this Guideline. Mr. Michael Halpern was the NUS Training Project Leader. The principal NUS Training authors were Sandra Baker, Michael Halpern, and Kathryn Kidd. Mr. Robert Walter, formerly of NUS Training, was also a contributor. Portions of this book are materials reprinted or adapted from NUS Training Corporation's *Procedure Writing Workshop Manual* (copyright 1990–1995).

CCPS also gratefully acknowledges the insight, thoughtful comments, and suggestions contributed by the following peer reviewers: Peter McGrath, Olin Corporation; Deric Crosby, PCR, Inc.; Peter Fletcher, Raytheon Engineers and Constructors, Inc.; J. Henry Blunt, Shell Oil Company; Thomas O. Gibson, The Dow Chemical Company; Rudolf Frey, The M. W. Kellogg Company; Glenn Damerell, 3M; Lester Wittenberg, AIChE/CCPS; Pamela Sutherland, Battelle; Tony R. Williams, JBF Associates, Inc.; John D. Snell, Occidental Chemical Corporation; and Leslie A. Scher, W. R. Grace & Co.

GLOSSARY

Accidental chemical release: An unintended or sudden release of chemical(s) from manufacturing, processing, handling, or on-site storage facilities to the air, water, or land.

Administrative procedures: Written instructions describing how to carry-out programs such as procedure writing and training programs.

Branching: A method used to send the reader to another location in the same procedure or to another procedure.

Checklist: A written list in which each item is marked off (or acknowledged on a computer screen) as its status is verified.

Critical: Relates to major environmental or safety risks.

Effective: Producing a definitive or desired result.

Emergency operating procedures (EOP): Written instructions that address actions to take to place a process in a safe and stable mode following a system upset.

Emergency Response Plan: A written plan required by governmental agencies which addresses actions to take in case of plant fire, explosion, or accidental chemical release of predetermined severity.

EPA: U.S. Environmental Protection Agency; Federal agency with environmental protection regulatory and enforcement authority.

Event: An identifiable occurrence related to equipment performance or human action, or an occurrence external to the system that causes (or could cause) dangerous system upset. The cause or contributor to an incident or accident.

Facility: Buildings, containers or equipment which contain a process.

Human factors: A discipline concerned with designing machines, operations, and work environments so that they match human capabilities, limitations, and needs. Includes any technical work (engineering, procedure writing, worker training, worker selection, etc.) related to the human factor in worker-machine systems.

Incident: An unplanned event or series of events and circumstances that may result in an undesirable consequence.

Job: A group of closely related tasks that achieve a specific goal or involve a common product, process, or service.

Maintenance procedures: Written instructions that address material control and maintenance practices needed to ensure system operability and integrity as well as maintenance, testing, and inspection frequency.

Management of Change: A system to identify, review, and approve all modifications to equipment, procedures, raw materials and processing conditions, other than "replacement in kind," prior to implementation.

Operating procedures: Written step-by-step instructions and associated information (cautions, notes, warnings, etc.) for safely performing a task within operating limits.

Operating emergency: An event (see definition) which can result in an accidental chemical release or injury if appropriate actions are not taken.

OSHA: Occupational Safety and Health Administration of the U. S. Department of Labor; Federal agency with safety and health regulatory and enforcement authorities for most U.S. industry and business.

Probability: The likelihood of occurrence of an event or an event sequence during an interval of time or the likelihood of the success or failure of an event on test or on demand.

Process: Any activity involving a chemical including any use, storage, manufacturing, handling, or on-site movement of such chemicals, or combination of these activities.

Procedure management system: System used to consistently control the development, maintenance, review, approval, and distribution of procedures to ensure that procedures are up-to-date, accessible, and accurate.

Process Hazards Analysis (PHA): An organized effort to identify and evaluate hazards associated with chemical processes and operations to enable their control. This review normally involves the use of qualitative techniques to identify and assess the significance of hazards. Conclusions and appropriate recommendations are developed. Occasionally, quantitative methods are used to help prioritize risk reduction.

Process safety: A discipline that focuses on the prevention of fires, explosions, and accidental chemical releases at chemical process facilities. Excludes classic worker health and safety issues involving work surfaces, ladders, protective equipment, etc.

Process Safety Management (PSM): A program or activity involving the application of management principles and analytical techniques to ensure the safety of chemical process facilities. Sometimes called process hazard management.

Safety: The expectation that a system does not, under defined conditions, lead to a state in which human life, economics or the environment is adversely affected.

System upset: A condition where any system operating parameter falls outside established safe operating limits, resulting, or likely to result, in unstable operation, operation outside of design limits, or potential release.

Task Analysis: A method for determining the detailed performance required of people and equipment and determining the effects of the surroundings, malfunctions, and other unexpected events on both.

Task: A meaningful unit of work with a measurable value that contains more than two steps and has a clearly defined starting and stopping point.

Task List: A list identifying and cataloging tasks performed by a job position or job area through research, analysis, and observation. Task List information may supply the base for the development of written procedures. A Task List can also be the product of a Task Analysis.

Validation: The act of testing the completeness and accuracy of a written procedure by observing a user perform or simulate the task.

1

INTRODUCTION TO EFFECTIVE PROCEDURE WRITING

1.1. Why Was This Book Written?

Pick up a dictionary and find the entry for *procedure*. One definition is: *a particular course or mode of action*. A simple statement, but to procedure users, writers, and managers concerned with process safety, the word *procedure* implies much more. The Center for Chemical Process Safety's approach to Process Safety Management relies on written procedures to document both hands-on operations and maintenance activities as well as to describe the Process Safety Management system itself. This book was written to help select and implement actions that will

- Help reduce accidents
- Promote safe, efficient operation and maintenance
- Improve quality, continuity, profitability, and cost control
- Build upon and record process experience
- Help maintain your competitive edge
- Reflect the best thinking on how to safely operate and manage your facility
- Comply with governmental regulations or industrial initiatives requiring written procedures
- Promote the idea that operating and maintenance procedures are vital plant components

The last item, treating procedures as if they were equipment (just like a pump, valve, reactor, or safety system), is fundamental for building a successful Process Safety Management system. Who would start up a new process without all of the pumps in place and tested? What craftsperson would tackle a pump seal replacement without the required tools and parts? By accepting this idea, that procedures are components, the concepts presented in this book will naturally fall into place.

Effectively written procedures can help to ensure the smooth startup of a new process or help solve a tricky problem. A large company with many sites can share procedures between plants with similar processes. A single site can gain by capturing temporary or seldom performed activities on paper for future use. Reinventing the wheel takes time that is better spent on productive activities.

Developing written procedures forces you to analyze your process, resulting in a better understanding of how to improve it. Safer, more efficient operations and maintenance follow, resulting in enhanced quality, decreased downtime, and lower costs.

Will effective procedures guarantee that your plant can avoid all accidents? No, they won't and they can't. However, as a component of your process safety management system, effective use of procedures can reduce the number of accidents caused by human error. Sound management practices encourage written procedures, and regulations now demand them for many processes. The purpose of this book is to make your procedure-writing efforts as productive as possible and to create accurate, effective procedures.

1.2. Book Objectives

There are three objectives for this book. Keep them in mind as you use this book. This book will

- Guide readers in developing effective written procedures to operate and maintain both large and small facilities
- Encourage development of procedure management systems that involve the users, emphasize control of the procedures, and require management and employee commitment to procedure use, accessibility, training, and management of change
- Help procedure writers produce effective, usable procedures that support safe and efficient operations and maintenance through improved communication

1.3. The Current Worldwide Trend Toward Procedures

Operating and maintenance procedures have become an international issue. Europe, Canada, and the United States have all issued or proposed regulations that require procedures for process safety purposes. Industry groups supporting offshore oil exploration have also recognized the value of procedures to improve process safety. The International Organization for Standardization (ISO) includes the control and use of procedures as a key element of a company's quality management system. These influences affect more companies each year as they seek a global market for their products.

In the United States, the impact of the Occupational Safety and Health Administration's (OSHA) Rule 29 CFR 1910.119 "Process Safety Management of Highly Hazardous Chemicals; Explosives and Blasting Agents; Final Rule" and/or the Environmental Protection Agency's (EPA) Proposed Rule 40 CFR Part 68 "Risk Management Programs for Chemical Accidental Release Prevention" is already being felt. Developing, training on, and maintaining the required written operating and maintenance procedures is a significant effort. This work represents a major and critical share of the total compliance effort these regulations demand. Your written procedures link the human component to the process.

Even before these regulations existed, groups such as the American Institute of Chemical Engineers' Center for Chemical Process Safety (CCPS), the Institution of Chemical Engineers (IChE), the Chemical Manufacturers Association (CMA), and the American Petroleum Institute (API) all promoted written procedures as a sound management practice. This practice, if firmly held, can lead to improved plant safety and product quality. Table 1-1 lists some organizations and documents that recommend or require procedures.

Whether a guideline, a regulation, or a recommended practice and whether for process safety, environmental, or quality reasons, a strong operating and maintenance procedure system will benefit your company.

1.4. Who Should Use This Book?

Anyone who needs procedures to ensure safe operations will benefit from using this book. Operating and maintenance procedures are important tools for many industries, especially those using or manufacturing chemicals. Refineries, commodity chemical plants, waste processors, plastics

TABLE 1-1
Organizations and Documents Recommending or Requiring Procedures

ORGANIZATION	DOCUMENT
AIChE Center for Chemical Process Safety (AIChE/CCPS)	*Guidelines for Technical Management of Chemical Process Safety*, AIChE/CCPS, New York, New York 1989.
American Petroleum Institute (API)	*API Recommended Practice 750, Management of Process Hazards*, American Petroleum Institute, Washington, DC, 1990.
Chemical Manufacturers Association (CMA)	*Responsible Care®: A Public Commitment*, Chemical Manufacturers Association, Washington, DC, 1988.
U.S. Environmental Protection Agency (EPA)	*Risk Management Programs for Chemical Accidental Release Prevention; Proposed Rule 40 CFR part 68*, Environmental Protection Agency, Washington, DC, 1993.
International Organization for Standardization (ISO) or in the United States of America American National Standards Institute (ANSI) 9000-1, 9001, 9002, 9003, 9004-1	*Quality Management and Quality Assurance Standards* (9000) ISO 9000—ISO 9004, *Quality Systems—Models for Quality Assurance* (9001-9003) *Quality Management and Quality System Elements—Guidelines* (9004) International Organization for Standardization, Geneva, Switzerland 1987.
US Occupational Safety and Health Administration, US Department of Labor (OSHA)	*Process Safety Management of Highly Hazardous Chemicals; Explosives and Blasting Agents; Final Rule 29 CFR 1910.119*, Occupational Safety and Health Administration, Washington, DC, 1992.
Organization for Economic Cooperation and Development(OECD)	*Workshop on Prevention of Accidents Involving Hazardous Substances Good Management Practice*, Discussion Document Relating to the Development of OECD Guideline for the Prevention of Accidents Involving Hazardous Substances, Organization for Economic Cooperation and Development, Berlin, 1989.
Official Journal of the European Communities	*Council Directive of 24 June 1982 (the Seveso Directive) on the Major-Accident Hazards of Certain Industrial Activities*, Official Journal of the European Communities, Berlin, 1989.
The World Bank	*Manual of Industrial Hazard Assessment Techniques*, The World Bank, Washington, DC 1985.

manufacturers, specialty chemical plants, and agricultural chemical processes—these all are obvious examples of processes that may use chemicals. Other, less obvious facilities with process safety and quality concerns include the food and beverage industry, pharmaceutical plants, pulp and paper plants, water and sewage treatment plants, utilities, household products plants, and many factories that manufacture everyday items.

Developing procedures is a team effort. The following functions at your facility will benefit from written procedures.

- **Operations and Maintenance**—Operators and maintenance personnel use procedures on a daily basis and are essential contributors to developing effective procedures. Their regular use of procedures can ensure revisions are made in a timely fashion.

- **Management**—Effective procedures capture the approved mode of operation and maintenance. Written procedures are the best way to communicate this information provided that there is regular review, comment, and response from the users. This book recommends administrative practices, including approval and review responsibilities and control methods.

- **Engineering**—Process hazard evaluations, process control improvements, turnarounds, and startups all benefit from effective procedures that accurately describe how to operate and maintain process equipment. This book presents the basic methods for ensuring that the procedures are effective and are reviewed for accuracy.

- **Safety**—Written operating and maintenance procedures provide information on hazards and the actions required to mitigate them. Lessons learned from incidents can also be included in the procedures when applicable. This book describes how to alert users of hazards before performing a task.

- **Environmental**—Using written procedures minimizes accidental chemical releases to the environment due to unsound operating and maintenance practices. If an incident occurs, the use of procedures should specify the proper response.

- **Training**—Written procedures are inextricably linked to operations and maintenance training. Well-written procedures provide a solid means of meeting the need to train workers on safe operations and maintenance. This book describes why procedures are important training tools.

Anyone given the responsibility for writing a new procedure or updating an existing one at your facility is a "procedure writer." Writers will benefit from this book by having a common basis for making decisions concerning style, content, and level of detail. Procedure users may include operators, maintenance craftpersons, Process Hazards Analysis (PHA) team members—potentially anyone in the plant. Each can benefit from this book by gaining an understanding of how procedures are written. This knowledge will help users gain the confidence needed to recommend changes in procedures when required.

1.5. Where Do You Go From Here?

Even if your procedures aren't written, they do exist. They exist in the experience you and your co-workers have gained over the years. That experience, combined with this book, could be all you need to begin writing your procedures.

Chapter 2 examines several voluntary guidelines and regulatory requirements dealing with written operating and maintenance procedures. Understanding the elements found in these guidelines and regulations helps to produce well-written, usable procedures that will assist you in reaching your process safety, environmental, and quality goals.

Chapter 3 describes the importance of written procedures and how to design a system for guiding procedure development and revisions. This procedure management system works with Management of Change to ensure procedures are selected, developed, and reviewed consistently using a planned framework that allows their easy use and maintenance.

Chapter 4 addresses procedure writing, including format selection, the types of references and resources needed, and the human factors of procedures. It describes how to begin writing, how to address technical data, and special considerations for maintenance procedures and batch processes.

Chapter 5 provides a starting point for developing a list of elements for writing effective procedures to meet your facility's needs. Due to your operations, maintenance, and regulatory concerns, you may want to add facility-specific procedure criteria to this list or delete criteria not applicable to your situation.

Chapter 6 discusses several factors to consider when identifying and writing for abnormal or emergency conditions. We define an operating emergency as an event which can result in an accidental chemical release or injury if appropriate actions are not taken. Such actions are described in Emergency Operating Procedures (EOP).

Chapter 7 presents the basic requirements to access, review, approve, control, and maintain procedures. It also discusses the wide range of choices you have in selecting the level of control your plant needs.

Chapter 8 summarizes why operating and maintenance procedures are a critical part of Process Safety Management, the typical costs associated with procedure development, and the return on investment.

Figure 1-1. The importance of effective procedures.
Reprinted by permission of Chronicle Features, San Francisco, CA.

2

PROCESS SAFETY, ENVIRONMENTAL, AND QUALITY CONSIDERATIONS

2.1. Purpose

In this chapter we will examine several voluntary guidelines and regulatory requirements dealing with written operating and maintenance procedures. Understanding the elements found in these guidelines and regulations helps to produce effective procedures that will assist you in reaching your process safety, environmental, and quality goals.

2.2. Understanding the Guidelines and Regulations

There are many reasons why you need to understand the requirements and recommendations of the various voluntary initiatives and regulations covering written operating and maintenance procedures. Written procedures

- Help maintain your competitive edge
- Make good business sense from a process safety, quality, and environmental point of view
- Reflect the best thinking on how to safely operate and manage your facility
- May be needed to comply with one or more governmental regulations requiring written procedures
- May be required for a desired certification

Figure 2-1. Safety, environmental, and quality responsibilities are interrelated

As demonstrated by common elements of the guidelines and regulations, quality, process safety, and environmental responsibility are interrelated. If a facility is free of accidents, quality is easier to achieve or maintain. A safe, quality-conscious facility will likely be an environmentally responsible facility. The elements of process safety, including effective, controlled procedures, combine to achieve the overall goal of a safe, environmentally responsible, quality operation.

In this chapter we will discuss guidelines, recommendations, and regulations addressing procedures. Their common elements may affect the way you manage your procedures. Many voluntary guidelines and recommendations served as the basis for subsequent regulations. Appendix A, *Selected Initiatives, Consensus Codes, and Regulations Affecting Procedures* lists some of these guidelines, recommendations, and regulations.

2.3. Voluntary Guidelines

In response to highly publicized incidents such as Seveso (Italy), Flixborough (UK) Three Mile Island (US), and Bhopal (India), process safety has become a global concern. Efforts to develop Process Safety Management in an attempt to reduce accidents began in Great Britain, and have since spread around the world. A result of this growing concern with Process Safety Management is recognition of the importance written operating and maintenance procedures can play in the prevention of accidents.

In response to process safety concerns, several organizations developed process safety guidelines. The guidelines address procedures and define the elements that go into developing and controlling written procedures and procedure management systems. These guidelines served as the basis for regulations. For example, in 1988, Organization Resources Counselors (ORC), Incorporated submitted *Recommendations for Process Hazards Management of Substances with Catastrophic Potential* to the U.S. Occupational Safety and Health Administration (OSHA). These recommendations, based on the input of a Process Hazard Management Task Force established by ORC, served as a basis for OSHA's Process Safety Management (PSM) Rule.

In 1989, the Organization for Economic Cooperation and Development held the "Prevention of Accidents Involving Hazardous Substances" workshop in Berlin, Germany. Acknowledging the importance of written procedures for the prevention of accidents involving hazardous substances, the workshop's written report (paragraph D.2) recommended that "each location should develop written job and process operating procedures sufficient to keep operating conditions within the design intent and maintain the integrity of the plant." [1]

As another example, the International Labor Office's "Prevention of Major Industrial Hazards" explicitly states that "good operational instructions and sound practices should be provided and enforced by the works management." [2]

Organizations such as the World Bank also recommend written procedures relating to administrative, accounting, and financial matters for facilities involved in the use and production of hazardous materials. The techniques described in this guideline can also be used to develop these types of procedures.

2.3.1. Center for Chemical Process Safety of the American Institute of Chemical Engineers

The American Institute of Chemical Engineers (AIChE) has been a leader in promoting process safety for many years. In recognition of the need for Process Safety Management to help prevent catastrophic chemical accidents, the AIChE established the Center for Chemical Process Safety (CCPS) in 1985. In 1989, the CCPS published *Guidelines for Technical Management of Chemical Process Safety*, one of the first titles in a series addressing process safety. This guideline addresses the importance of written procedures as part of an overall Process Safety Management system.

TABLE 2-1
Important Points to Be Addressed as Summarized from CCPS "Guidelines for Technical Management of Chemical Process Safety"

Procedures should identify

- Tasks to be performed by the operator
- Instrument readings and samples to be taken
- Conditions to be maintained
- Safety precautions
- Safe operating limits for critical parameters
- Critical operating parameters and instruments
- Results of operating beyond safe limits
- Corrective and emergency actions

Procedures should

- Be accurate
- Be understandable
- Use familiar language
- Include input from process and design engineers and operations and maintenance personnel
- Reflect how operations are actually performed
- Be thoroughly documented
- Be dated and/or have a revision number on every page
- Be reviewed and updated at regular intervals to capture procedural, equipment, critical operating parameter, software, and process changes
- Be approved

Users should

- Be informed of all changes
- Be familiar with critical operating procedures

It also deals with factors relating to the development and use of written procedures. Table 2-1 gives an overview of the major points discussed in the CCPS guideline. These points should be addressed when developing procedure management systems and written procedures.

2.3.2. Chemical Manufacturers Association

Another industry initiative, the Chemical Manufacturers Association (CMA) Responsible Care® Process Safety Code of Management Practices, refers to operating procedures by noting the need for "current, complete documentation of process design, operating parameters, and *procedures*"[3] (emphasis added).

2.3.3. American Petroleum Institute

The American Petroleum Institute's (API) Recommended Practice (RP) 750 was issued in 1990 in response to the growing concern for process safety and the prevention of catastrophic releases of toxic or explosive materials. As shown in Appendix B, *Common Points of the API RP 750, OSHA's PSM Rule 29 CFR 1910.119,* and *EPA 40 CFR Proposed Rule Part 68,* major elements of API's recommendations are reflected in both the OSHA and Environmental Protection Agency (EPA) regulations.

2.4. Governmental Regulations

In an attempt to further reduce the number and effect of catastrophic chemical incidents, government regulatory agencies have defined and enforced many of the elements of process safety put forth by the voluntary guidelines. Two influential regulations are from OSHA and EPA.

2.4.1. United States Occupational Safety and Health Administration

OSHA's PSM Rule requires written procedures for facilities with specified quantities of identified hazardous chemicals. Paragraph (f) of this regulation directly addresses the need for operating procedures and specifies requirements for those procedures, including the contents and the categories of operations they should address. Paragraph (j), Mechanical Integrity, requires employers covered by the regulation to establish and implement written procedures to maintain the ongoing integrity of process equipment. This regulation is "a process safety standard in order to protect employees from hazards associated with accidental releases of highly hazardous chemicals in the workplace." [4]

2.4.2. United States Environmental Protection Agency

The EPA's proposed rule 40 CFR Part 68, "Risk Management Programs (RMP) for Chemical Accidental Release Prevention; Rule Subsection 28 Prevention Program—Standard Operating Procedures," also identifies the need for operating procedures in language parallel to OSHA's PSM Rule while reflecting EPA's concern with off-site consequences. The goal of the proposed RMP rule is to "prevent accidental releases of regulated substances to the air and to minimize the consequences of releases by focusing preventive measures on those chemicals that pose the greatest risk." [5]

2.5. Quality Considerations

The guidelines and regulations cited so far have focused on process safety. However, certification and award programs such as the International Organization for Standardization (ISO) Standard 9002, "Quality Systems—Model for Quality Assurance in Production and Installation," place heavy emphasis on quality management. Written procedures and document control are important factors for ensuring and maintaining quality, reliability, and consistency.

The ISO Standard 9002 is a major global influence on procedures and document control. It identifies the need for production and installation instructions where the absence of such instructions would adversely affect quality. Although this is a general statement, it is difficult to think of a task worth doing that does not or cannot potentially affect quality as well as safety. The ISO standard addresses procedures in general by requiring that all documentation involved with quality be controlled. Table 2-2 outlines the major points of ISO 9002 dealing with documentation and procedures.

TABLE 2-2
A Brief Summary of ISO 9002 References to Procedures

4.4. Document control—Establish and maintain procedures to control all documents and data that relate to requirements of standard. Documents shall be reviewed and approved for adequacy by authorized personnel prior to issue. (a) Appropriate documents available at all locations where operations essential to the effective functioning of a quality system are performed; (b) Obsolete documents promptly removed from all points of issue or use.
4.4.2. Document changes/modifications —Changes shall be reviewed and approved —Nature of change identified in document/attachments —Procedure in place to preclude use of non-applicable procedures —Documents reissued (revised) after practical number of changes made
4.8.1. (a) Process control—Establish instructions defining manner of production and installation where absence of instructions would adversely affect quality
4.12. Control of nonconforming product—Establish and maintain procedures to ensure nonconforming product not used or installed
4.18. Statistical techniques—where appropriate, establish procedures identifying statistical techniques for verifying acceptability of process and products

2.6. Some Elements of Effective Procedures and Procedure Management Systems

These consensus codes and regulations share common elements that can guide you in developing and evaluating procedures. These elements are shown in Table 2-3. In general, all the guidelines and regulations require current, complete documentation of process safety information including:

- Process chemistry
- Process design and equipment specifications
- Process operating procedures
- Operating limits

You will find that the common elements discussed in this section will help make your procedure management system more effective.

2.6.1. Accessibility

Procedure users must be able to quickly and easily obtain current, approved procedures to prepare for and perform their jobs. Needed procedures must be readily accessible—available—at all times. Procedures may be available as printed (hard-copy) documents, they may be viewed on computer screens, or they may be printed, as needed, from electronic files. The current, approved procedures must be available to ensure that only up-to-date procedures are used to perform operations and maintenance tasks.

2.6.2. Clarity

In addition to being readily available, procedures must be clear. They must be written concisely in a straightforward manner and must consider both the difficulty and importance of the task(s) being described. They must also consider the skills, experience level, and needs of the user. If the user does not understand a procedure, or does not have confidence in its accuracy, the procedure will most likely not be used or it will be used incorrectly. Procedure training will foster understanding and use of procedures.

2.6.3. Control

Many of the guidelines and regulations cited address the need to ensure procedures are current and accurate. This means that a procedure management system should be in place to implement and guide the development,

TABLE 2-3
List of Procedure Elements, Common Regulations, and Consensus Codes[a]

Procedure Elements	CCPS (Safety)	API (Safety)	OSHA (Safety)	EPA (Environment)	ISO (Quality)
Procedures accessible		✔	✔	✔	
Procedures contain clear instructions	✔	✔	✔	✔	
Procedures address steps for each operations phase	✔	✔	✔	✔	
Procedures reviewed/ approved before issue	✔	✔	✔	✔	✔
Procedure approved by responsible party	✔	✔	✔	✔	✔
Procedures reviewed for accuracy/adequacy	✔		✔	✔	✔
Procedure changes documented	✔		✔	✔	✔
Method in place for control/removal of obsolete procedures	✔		✔	✔	
Periodic review of operating procedures		✔	✔	✔	
Annual Certification confirming that operating procedures accurately reflect current practices		✔	✔	✔	
Safe operating limits	✔	✔	✔	✔	
Deviations and consequences	✔	✔	✔	✔	✔
Safety & health considerations	✔	✔	✔	✔	
Safety systems		✔	✔	✔	
Emergency actions			✔	✔	
Safe work practices			✔	✔	
Training		✔	✔	✔	
Procedures maintaining mechanical integrity		✔	✔	✔	
Employee input	✔		✔		

[a]See Appendix A, Selected Initiatives, Consensus Codes, and Regulations Affecting Procedures.

review, approval, distribution, accessibility, and updating of procedures. (See Chapter 3, *How to Design an Operating and Maintenance Procedure Management System.*)

As mentioned in Chapter 1, we should treat procedures with the same respect as we do equipment and process materials. They are a major investment. Revisions or modifications to procedures should be analyzed, tracked, and approved in the same manner as mechanical or technological changes. To those familiar with the elements of process safety, this is part of "Management of Change" or "Change Control." The CCPS *Guidelines for Technical Management of Chemical Process Safety* also recommends documenting all procedure changes. If your facility is considering ISO 9000 series accreditation, procedure control will be a key element.

2.6.4. Reviews and Audits

To ensure that procedures are accurate and reflect current practices, they must be periodically reviewed. Revisions caused by changes or improvements in equipment, process technology, standard practice, or facility status must be incorporated as they occur. This is a function of your procedure management system. The effect of changes in environmental and safety regulations on procedures must not be overlooked. A Management of Change system directly supports and controls these revisions.

2.6.5. Employee Input

Your co-workers are important resources for developing, reviewing, and updating procedures. CCPS recommends that input be included from operating and maintenance personnel as well as process and design engineers. OSHA requires employee participation in all phases of Process Safety Management. Again, this makes good sense; after all, employees at all levels are a critical element in maintaining process safety.

2.7. Additional Considerations

The following requirements of the OSHA and proposed EPA regulations are topics that deserve consideration even if your facility is not directly affected by these regulations. Industry guidelines and initiatives also promote these elements. Appendix C to OSHA 29 CFR 1910.119, *Com-*

pliance Guidelines and Recommendations for Process Safety Management (Nonmandatory), contains additional helpful advice on procedures and should be reviewed as a useful resource when you develop your procedure management system.

2.7.1. Training

In addition to helping to ensure quality, procedures can be used as training resources and on-the-job training aids or scripts. Effective procedures are an important factor in maintaining consistency and ensuring that everyone receives the same baseline level of training and information. Written procedures may be used as the primary training device, serving as a guide to train workers to perform their jobs. They may also serve the same function as part of a more extensive training structure.

2.7.2. Safety and Health Considerations

Procedures should identify the hazards presented by the process. Procedures should also state precautions necessary to prevent accidental chemical release, exposure, and injury. Process safety information is an important resource in developing procedures. Using this information ensures that the known hazards are addressed properly.

2.7.3. Safe Work Practices

Practices that concern process and personnel safety such as steps for controlling hazardous energy, vessel entry, line opening, confined space entry, and flame or spark-producing work (hot work) should be addressed in procedures either as steps or as references.

2.7.4. Operating Limits

Including safe operating limits, the consequences of deviating from those ranges, and how to correct/avoid those deviations in procedures helps to prevent accidental chemical release, exposure, and injury. This information enhances the users' understanding of the process, its limitations, and how to run it safely.

2.8. Conclusion

Although OSHA's PSM Rule, EPA's Proposed Rule, and API's Recommended Practice 750 identify categories of procedures to be written, they do not clearly define those categories. Various regional and local authorities may have additional procedure requirements. Based on the type(s) of process(es) at your facility, such as batch or continuous, you will have to decide which categories are appropriate and write your procedures accordingly.

Table 2-3, *List of Procedure Elements*, lists the elements related to procedures and procedure control and indicates which guideline(s) and/or regulation(s) identify the element. It is intended to be used as a checklist to help you determine which elements are applicable or required for your facility. Methods for addressing the elements required to design an effective procedure management system are discussed in Chapter 3.

The various guidelines, initiatives, and regulations reflect an emerging consensus that ties written procedures to safety, environmental, and quality considerations. Understanding these requirements and recommendations will help ensure that your procedures contain the critical information to run your process safely and effectively.

Endnotes

1. Organization for Economic Cooperation and Development (OECD). *Discussion Document Relating to the Development of OECD Guidance for the Prevention of Accidents Involving Hazardous Substances* from the Workshop on Prevention of Accidents Involving Hazardous Substances Good Management Practices. Berlin, Germany, 1989.
2. International Labor Office (ILO). *ILO Code of Practice for the Prevention of Major Industrial Accidents*. Geneva, Switzerland, 1991.
3. Chemical Manufacturers Association. *Responsible Care®: Process Safety Code of Management Practices.* Washington, DC, 1990.
4. United States Department of Labor, Occupational Safety and Health Administration. *Federal Register* Volume 57, Number 36, pp. 6356–6417: "29 CFR Part 1910.119: Process Safety Management of Highly Hazardous Chemicals"; "Explosives and Blasting Agents," Final Rule. Washington, DC, February 1992.
5. United States Environmental Protection Agency. *Federal Register* Volume 58, Number 201, pp. 54190–54219, "40 CFR Part 68: Risk Management Programs for Chemical Accidental Release Prevention." Washington, DC, October 1993.

3

HOW TO DESIGN AN OPERATING AND MAINTENANCE PROCEDURE MANAGEMENT SYSTEM

3.1. Purpose

This chapter describes the importance of written procedures and how to design a system for guiding procedure development and revisions. This procedure management system works in conjunction with Management of Change to ensure procedures are selected, developed, reviewed, and approved consistently using a planned framework that allows their easy use and maintenance.

3.2. The Importance of Written Procedures

Effective operating and maintenance procedures provide a win–win opportunity for both facilities and individual employees. Here are just a few of the recognized benefits of effective, written operating and maintenance procedures. They

- Provide a record of approved, safe operating and maintenance practices
- Provide consistent information to all users
- Remove guesswork
- Support employee experience and knowledge

- Enhance employee performance
- Document and build upon your facility's experience and practices
- Assist in adhering with industry initiatives and regulations
- Lead to more efficient operations
- Provide the tools for an effective training program

Developing effective operating and maintenance procedures requires commitment and resources. Part of the development includes designing a procedure management system. The benefits of the investment in your procedure program, listed above, will offset your initial costs.

3.3. Elements of a Comprehensive Procedure Management System

The purpose of your procedure management system is to define how you will consistently develop, review, approve, implement, maintain, and control your procedures. Describing your procedure management system in a written, administrative procedure is a recognized and effective way to clearly establish those instructions. The procedure management system is based on your available resources (human, technical, and financial) and the hazard potential of the processes involved. Remember, your objective is to achieve a safe, environmentally responsible, quality operation.

Recall our analogy of a procedure as a fundamental piece of equipment similar to a pump. This analogy shows the importance of the procedure management system. Assume that you have been given the task of specifying and purchasing a pump. You would determine the type of pump you need and its flow and pressure ratings. You would then choose the most effective pump available. You should treat your operating and maintenance procedures in the same manner. The time spent designing an effective procedure management system is just as critical to success as the time that was spent purchasing the correct pump.

You would not think twice about performing the necessary maintenance on a pump. It is an investment and vital to production. Procedures are also an investment and are vital to production. They must be maintained to ensure they remain functional and effective.

The following is a general outline for developing procedures. A procedure project may be initiated because there are missing or out-of-date procedures. The first step in any procedure project is to design and implement a procedure management system, or to evaluate your current system and improve it as necessary. Each of the steps in this general outline is discussed in further detail in this chapter.

1. Determine your procedure management system requirements.
2. Compare your current practices against your procedure management system requirements. Do improvements need to be made?
3. Identify the concepts in this book that will help you meet your procedure management system requirements.
4. Identify your available human, financial, and technical resources.
5. Design and implement your procedure management system.
6. Identify the operating and maintenance procedures that need to be written or revised.
7. Implement the procedure project by assigning personnel, scheduling development, writing procedures, and putting them in use.

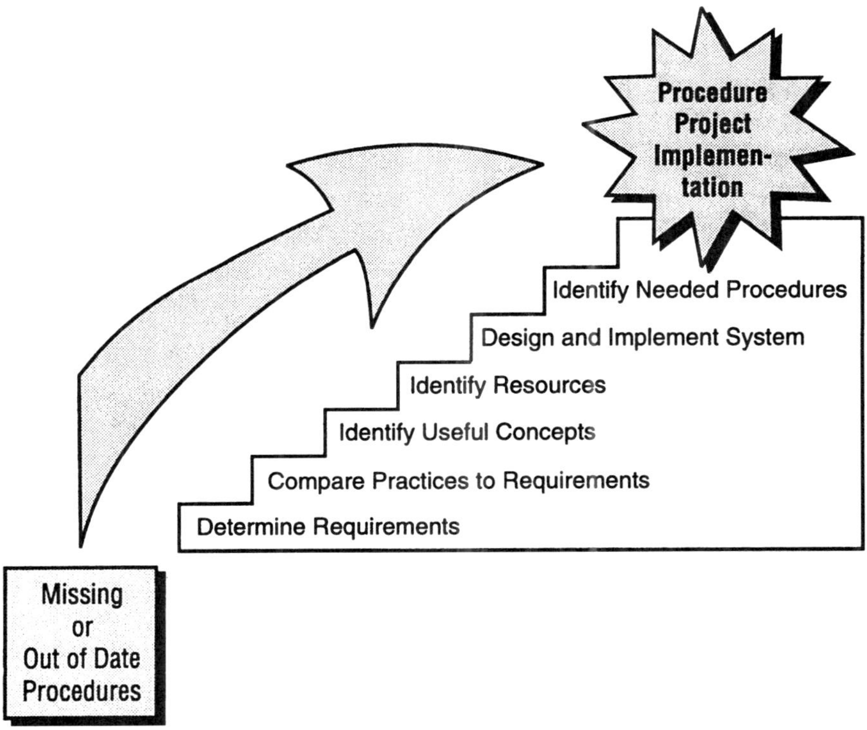

Figure 3-1. The steps to implementing a procedure project.

3.4. Determining Procedure Management System Requirements

A procedure management system should be based upon company requirements to satisfy process safety, quality, and environmental needs. The more clearly you define your system requirements, the easier it will be to design the system.

Chapter 2 discussed various initiatives and regulations that may be applicable and that help determine what procedure management system must deliver. The checklist presented as Figure 3-2 addresses some common requirements and can be customized to meet your facility's needs.

This checklist provides a method for producing and maintaining procedures which

- Accurately describe safe, current operating or maintenance practices
- Prevent accidental chemical releases that may injure employees, the public, or the environment
- Describe how product quality will be maintained

3.5. Evaluating Your Current Practices

Once you have identified the system requirements, you should evaluate your current practices against these requirements by identifying, at a minimum,

- The status of your procedure management system
  —Do you have any policies, documents, or standards that govern procedure development or control?
  —Do these policies, documents, or standards work effectively to ensure development and maintenance of effective, accurate procedures?
  —Do your current procedures comply with your policies and standards?
- The status of your procedures
  —Do you have any written procedures?
  —Do they include important safety information?
  —Are they accessible?
  —Do your procedures reflect the operating or maintenance tasks as they are actually performed?
- The effectiveness of your procedures
  —If there are written procedures, do they satisfy your safety, quality, and environmental needs?

PROCEDURE MANAGEMENT SYSTEM DESIGN CHECKLIST

Procedure Development Preparation (This Chapter)

- ☐ Assign responsibility for developing your procedures
- ☐ Determine how you will identify what procedures need to be written
- ☐ Set schedule requirements for developing the procedures
- ☐ Document the procedure management system

Procedure Development (See Chapter 4)

- ☐ Develop your procedures so that they accurately reflect operations and maintenance
- ☐ Develop the procedures so that they will be used
- ☐ Select a style and format to ensure that your procedures are consistently and effectively written
- ☐ Establish clear definitions for the different categories of operating and maintenance procedures
- ☐ Decide how users access the procedures

Procedure Review and Approval (See Chapters 5 and 7)

- ☐ Establish a procedure review cycle
- ☐ Determine how you will measure procedure accuracy and effectiveness
- ☐ Determine who will approve your procedures
- ☐ Review procedures as a whole to identify any information gaps

Procedure Control (See Chapter 7)

- ☐ Establish a way to prevent obsolete procedures from being used
- ☐ Ensure that any changes to the operating or maintenance technology or method of operation will be incorporated into the procedures. This must be accomplished through a Management of Change system
- ☐ Conduct periodic review of procedure management system and procedures
- ☐ Revise the procedure management system as necessary

Figure 3-2. Checklist for designing a procedure management system.

—Are your procedures written so that they can be easily followed and understood by the intended users?

—Are your written procedures actually followed by the operators or mechanics?

—Are your procedures written so that they meet the needs of your training program?

3.6. Identifying Your Resources

While designing your procedure management system, you must determine the resources that are available to you. There are many resources to consider, but they can be divided into four basic categories: people, capital, existing information, and computer hardware. Some common questions that relate to resources are

- Who will write the procedures?
- Who will review the procedures?
- Who will revise the procedures?
- Who will carry out the various responsibilities outlined in the procedure management system?
- What computer resources are available to the procedure writers and those managing the procedures?
- Who will distribute revised procedures to replace old ones whether as hard copies or on a computer?
- How will revised procedures be distributed?
- What information already exists that can be used to develop or revise procedures?

3.7. Designing and Implementing Your Procedure Management System

Designing your procedure management system involves deciding how you will use your resources to meet your system requirements. (Refer to Figure 3-2, the checklist on the preceding page.) You must determine how you will meet each requirement based upon your resources. This must be clearly described in your procedure management system. For example, one of the procedure management system requirements is to ensure that the procedures can be accessed by the procedure users. Some ways to achieve this requirement are to

- Use hard copies of the procedures contained in designated procedure manuals
- View the procedures at a computer terminal
- Print out hard copies of the procedures from computer terminals to take out into the plant

Every facility may have a different method for ensuring that employees can access the procedures, but that method needs to be defined.

There are three aspects to maintaining up-to-date procedures. First, you must recognize when a new procedure or procedure revision is needed. Second, you must develop or revise the procedure and implement it. Third, you must make certain that only current procedures are in use. (See Chapter 7, *Procedure Control*.)

Part of designing your procedure management system is to consider the interaction between that system and your Management of Change (MOC) system. The MOC system should identify the need for revisions to procedures. For example, replacing a pump with exactly the same pump ("replacement in kind") requires no procedure revisions. However, replacing a pump with a new pump that has additional or fewer auxiliary systems requires updating operating and maintenance procedures to reflect this change. Your procedure management system will detail how the development or revision of a procedure should be conducted.

A request for a procedure change could also trigger the MOC system. For example, to shorten batch cycle time, a change in the sequence of procedure steps is proposed. This would require an MOC evaluation for process safety, environmental, and quality impact. If the request is approved, the procedure should then be developed or revised according to the standards set in your procedure management system.

Once you have designed your procedure management system, the system should be reviewed by all necessary personnel, revised as necessary to ensure its effectiveness, and then approved. Since the system is a procedure itself, it should undergo the same evaluation and approval as the procedures it governs.

3.8. How to Determine Which Procedures to Write

Most of the work at your facility is guided by some form of procedure—sometimes documented, sometimes not. Here is a list of procedures you may find:

- Administrative Procedures
  - —Process Safety
  - —Training
  - —Environmental Control and Reporting
  - —Quality
  - —Finance
  - —Human Resources

- Operating Procedures (Process Area)
  - —Startup
  - —Shutdown
  - —Normal operating procedure
  - —Temporary
  - —Emergency
- Maintenance Procedures
  - —Mechanical
  - —Electrical
  - —Utilities
  - —Calibration/Testing
- Safety Procedures (Safe Work Practices)
  - —Lockout/Tagout
  - —Confined Space Entry
  - —Hot Work
  - —Line Breaking
  - —Electrical
- Engineering Procedures, Practices, and Standards

Once you have defined the type of procedures you are writing, one of the greatest challenges is to determine which tasks require written procedures. A task is defined as a measurable performance consisting of more than two actions with a clearly defined starting and stopping point. A job includes many related tasks. For example, being a gardener is a job. A gardener performs the tasks depicted in Figure 3-3, the flow chart below. Each of these tasks could be described in a written procedure.

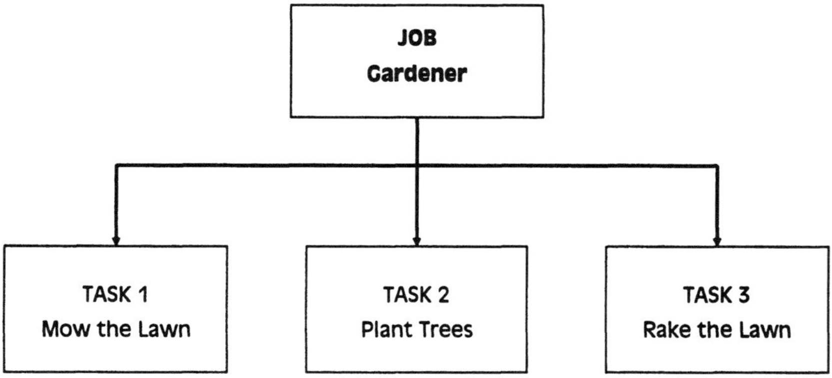

Figure 3-3. Job and task flow chart.

To determine which procedures must be written for a given job, you must first determine the tasks performed in that job. Although there are many different ways to identify tasks (refer to Appendix C, *How to Determine Tasks That Require Written Procedures*), most of these methods have six common steps:

1. Determine why you are writing procedures—safety, environmental responsibility, and/or quality. This will help you to identify the jobs that you want to evaluate.
2. Select a job to be evaluated.
3. Assemble and review existing material related to that job. Develop a preliminary list of tasks based on the compiled information.
4. Prepare the task list:
   a. Observe the actual job performance, recording the tasks that are performed. It is important to observe more than one person performing the job. For example, observe an operator on each shift. Additionally, you should interview the personnel to determine if there are any tasks that they perform infrequently that you might not have observed.
   b. Compare your observations against your preliminary task list to ensure you have recorded all of the tasks.
   c. If possible, route the task list to other personnel who perform the same job for additional input. Revise the list based on their input to develop the final task list.
   d. Arrange tasks in a logical order, such as the order in which they are performed, alphabetically, or by similarity. This will increase the ease of using and reviewing the task list.
5. Evaluate each task to determine if the task will require a written procedure. Factors that determine if a task requires a written procedure can be the frequency, criticality, and complexity of performing the task. Other factors can include regulatory requirements such as the OSHA 1910.119 Process Safety Management Rule procedure requirements. For example, starting a simple pump may not require a written procedure because it is a very simple task that people are trained to do from memory. However, starting a complicated pump with many auxiliary systems or a pump that is critical to safety may require a written procedure to ensure the pump is always started correctly.
6. Once you have identified which tasks must have written procedures, you are ready to begin your procedure writing effort.

Observing actual job performance may not always be possible. For example, if a facility is under construction, you may not have the opportunity to observe actual job performance. However, you may be able to examine technical information from another similar facility. You can conduct a systematic discussion involving, at minimum, the design engineer and operators and maintenance personnel with experience at similar facilities to determine the task list. Tasks are identified by talking through, instead of walking through, the operation of the unit using all relevant technical information and experienced personnel.

3.9. Implementing a Procedure Project

After you have identified the procedures that need to be written, you are ready to write them. Writing procedures is discussed in detail in Chapter 4. Developing a procedure schedule and assigning personnel will be based upon your deadlines and resources available.

3.10. Procedure Training

An important part of implementing procedures is training the personnel on the procedures. The objective of this training is to ensure that people know how to safely perform their job. The more closely the training reflects actual job performance, the more likely you are to achieve that objective. Effective procedures provide a clear definition of safe job performance. Therefore, procedures can be used to train personnel and to evaluate their ability to perform the job safely.

The first step in providing effective training is to make the training objectives clear. How often have you been to training where the objectives weren't clear and you asked yourself:

- What am I expected to learn?
- Why do I have to learn it?
- What will I be able to do or understand at the end of the training?

During on-the-job training, the answers to these questions should be apparent because effective procedures spell out what the user should be able to do. That is why procedures are a critical part of on-the-job training.

Once you have trained employees on a procedure, you want to know if they can actually perform it safely. Three items must be defined before evaluating a trainee's performance. What are the

- Conditions under which the trainee will perform the procedure?
- Actions that the trainee must perform?

- Standards against which the trainees will be evaluated to determine if they have successfully passed the evaluation?

The conditions under which the trainees will perform the procedure should closely resemble the conditions under which the employees will usually perform the procedure. For example, during their evaluation:

- Will they be allowed to use reference materials such as the procedure?
- Will they actually perform the task, or will they simulate the performance (often used when evaluating emergency procedures)?
- Will they have to perform the procedure from memory?

The actions the trainees must perform are the steps of the procedure. At a minimum, the standard that they must satisfy is to successfully perform the procedure as written. However, there may be additional performance standards, such as

- Was the procedure completed within the allotted period of time?
- Did the trainees identify all safety equipment?

Clearly defining the conditions, actions, and standards ensures that the trainee performance evaluation will be consistent and accurate. An example of a procedure performance evaluation form is included in Appendix D, *Procedure Performance Evaluation*.

In addition to training personnel on procedure content, it is important to train personnel on the procedure format. The procedure format is very similar to a roadmap. Before you can use a roadmap effectively and correctly, you must understand how to use the roadmap, what information it contains and where that information is located on the map.

3.11. Maintaining and Improving Your Procedure Management System

A procedure management system is successful only if effective procedures are being used. The system should be evaluated and updated to ensure that it is successful. Why do most procedure management systems fail?

- Lack of commitment to using procedures
- Lack of commitment to following the procedure management system
- Lack of understanding of the procedure management system
- Lack of procedure accessibility
- Lack of procedure training

- A defect in the procedure management system that allows changes to occur in process or technology without updating the procedures

For example, a facility launched an expensive effort to upgrade and develop procedures. Six months later, some of the procedures are no longer accurate. What happened? The procedure management system failed because of one or more of the above reasons.

Commitment to the procedure management system and the use of the procedures must come from all areas of your facility, but it is critical to obtain management and employee commitment and participation. Without everyone's commitment to the procedure management system, procedures will not be properly developed, maintained, or used. One way to foster commitment is to involve the people who will implement and use the system in its design. Training on the system increases their understanding, helps to ensure its consistent use, and increases the feeling of ownership in the system. A feeling of ownership encourages the users to provide feedback on the procedures and the procedure management system. Continuous improvement of the procedures and procedure management system is the result.

Lack of procedure accessibility or procedure training can also lead to failure of the procedure management system. If operating and maintenance personnel cannot access the procedures or are not properly trained to use them, the primary goal of the procedure management system will not be achieved. The most effective and accurate procedures add value only if they are used.

A deficiency in a procedure management system may not always be apparent. Periodic evaluation of the system, including the written procedures, will help to identify and correct deficiencies. Part of this evaluation should be the verification that the procedures reflect actual, approved plant practices. Procedure users are an excellent system evaluation resource. If your procedures are used frequently, the users will recognize when they no longer reflect current operations or if they begin to stray from the development guidelines.

3.12. Conclusion

A procedure management system guides procedure development and revisions. The product of this system is effective, consistent procedures which are easily maintained and controlled. Management and employee involvement is critical. Everyone's commitment to the procedure management system is essential to keeping the procedures up-to-date and effective.

4

WRITING OPERATING AND MAINTENANCE PROCEDURES

4.1. Purpose

This chapter addresses procedure writing, including format selection, the types of references and resources needed, and the human factors aspects of procedures. It describes how to

- Begin writing
- Address technical data
- Identify special considerations for maintenance procedures and batch processes

4.2. What Resources Do You Need Before You Begin Writing?

Before writing or revising a procedure, you need to consider and assemble at least the following sources of information:

- **Design, technical, and safety information.** Locate the design information for the unit or equipment. This can include Piping and Instrumentation Diagrams (P&ID) and Material Safety Data Sheets (MSDS) for all the materials involved in the process or for maintenance, equipment vendor manuals, process hazard analysis (PHA) results, safe work practice procedures, and job task lists. Additional sources of such data are lab, pilot plant, engineering, maintenance, folklore and long-term employee memories, contractors, and insurance records. Also, identify any updates or process upgrades that may have been done.

New projects should include preparation of operating procedures. If an engineering firm is used for the upgrade, they typically provide manuals containing this information as part of the contract.

- **Operating and equipment parameters, ranges, and limits.** Obtain a current list of the approved operating ranges and limits for the processes for which you are writing the procedures. These may be different from the original design specifications. Sample operating limits table formats are included in Appendix G.
- **Existing procedures.** All existing written procedures are helpful. These may be typed documents or handwritten guides that experienced operators or maintenance craftspersons kept for their own use. Remember, not all existing procedures are written. The procedure may exist only in the mind and experience of your personnel. This is why interviewing those personnel can provide valuable information.
- **Human resources.** The people familiar with the day-to-day operation of the process are valuable sources of information. They can provide the operational or maintenance information that makes your procedures unique and accurate. Even if you are writing procedures for a brand-new process, you can build a team that can provide input from their experience in working with similar processes.

4.3. What Do We Know About the Procedure Users?

Procedure writing is an exercise in the use of human factors. When you are writing procedures, there are two types of human factors to consider. The first is how the user actually uses the equipment to perform the task. The second factor is the relationship between the workers and the procedure they are using.

An open page format, one that is mostly open space, is less intimidating to the reader than a page cluttered with text. A full page of text often blurs into a gray square when a reader glances at it. Brief steps and a consistent format all lend themselves to proper use of the document.

Who will use the procedures you write? Operators? Maintenance personnel? Instrument Technicians? Engineering staff? Contractors and temporary employees? Health & Safety staff? The users of the procedure and the level of information needed to perform the task will determine how the procedure is written.

What experience or training does the intended user have? The users may be trainees unfamiliar with process work, experienced personnel learning a new or upgraded process, or highly skilled and experienced operators or craftspersons. Procedures must guide the least experienced, qualified worker yet still be useful to the experienced, qualified worker. In other words, if there is too little information, a novice might not be able to perform the task. Too much information, and even an experienced worker may discount the procedure as "too detailed and hard to follow."

An important part of Process Safety Management (PSM) is employee participation. By involving the employees in all phases of the PSM system, a facility develops a culture that promotes ongoing maintenance of that system. Writing and revising procedures provide many opportunities for this involvement. Employees who assist in these activities gain a sense of ownership and control over the tasks that make up their jobs. However, if experienced employees write the procedures, they should be trained in effective procedure writing. Here are some ways to encourage employee participation in preparing and maintaining effective procedures:

- Use experienced operators or maintenance personnel as subject matter experts in procedure development
- Use experienced employees as procedure writers
- Provide the procedure users an opportunity to review the procedures
- Implement a procedure management system that encourages any user to raise an issue or recommend improvements to an approved procedure

4.4. Considerations for Effective Procedures

An effective procedure has completeness and accuracy, appropriate level of detail, conciseness, consistent presentation, and administrative control.

4.4.1. Completeness and Accuracy

Completeness and accuracy are difficult elements for the writer to accomplish and for the reviewer to evaluate. They

- Depend on thorough research and analysis of the operation during the procedure development stage and a detailed review of the completed procedure by knowledgeable and responsible plant staff before approval

- Ensure that the procedure's goal is achieved and all conditions are satisfied

Completeness is not a function of the procedure's length or level of detail. Rather, it is a function of whether a procedure has enough information for the user to perform the task safely and correctly.

One way to test for completeness and accuracy is to have a typical user simulate or perform tasks using the written procedure. This may be a dry run, water batch, simulation, or actual use.

4.4.2. Appropriate Level of Detail

The level of detail is based on the responsibilities, training, experience level, and capabilities of the intended users. Level of detail also is determined by the criticality and potential hazards of the operation and ease or frequency of performance. Proper level of detail contributes to ease of use and comprehension. Care should be taken to ensure the procedure does not become cumbersome, thereby affecting its effectiveness. You have included the proper level of detail when the least experienced, trained user can safely perform the procedure as written.

4.4.3. Conciseness

Conciseness demands eliminating detail and language that do not contribute to work performance, safety, or quality; include only "need-to-know," and omit "nice-to-know" information. "Need to know" means just the information required to safely and efficiently perform the task. For example, when measuring pH an operator must know how to use pH paper properly and the specified pH limits. It is nice for the operator to know how the pH paper works.

4.4.4. Consistent Presentation

This element ensures that the procedure is readily comprehensible. It demands the use of

- A consistent terminology for naming components and operations
- A standard, effective format and page layout
- A vocabulary and sentence structure suitable for the intended user

4.4.5. *Administrative Control*

All procedures need to be reviewed thoroughly before use and periodically thereafter. This ensures incorporation of plant or process changes and compatibility with current plant status and plant requirements. The responsibilities for editorial review, technical review, and document control must be defined. This is discussed in detail in Chapter 7, *Procedure Control.*

4.5. Importance of Procedure Format

Procedure writing is straightforward if you prepare properly and follow a well-thought-out and functional format. The format should guide the user to the final goal or destination. Additionally, the format should guide the writer during the development of the procedure. A procedure written using standard format is like a road map.

When traveling to a new location, a traveler uses a road map

- Before leaving, in order to know what to expect in terms of traffic and the types of roads (city streets, local roads, or superhighways), and to get a feel for the number of turns, the distance, and the estimated duration of the journey.
- While driving, to check that the proper turns were made, to look ahead for rest or fuel stops, and to estimate progress.

If traveling to a location again and again, the map may not be needed as often. But if trouble arises (detours, construction, or a traffic jam) or if it has been some time since the last trip, the map can be used to solve the problem or answer a question. Procedures can be used in the same way.

4.5.1. *Page Layout*

Procedure formats vary according to user needs, acting to guide the reader through the procedure to extract and use the information in an efficient manner. The way you present the procedure steps and words on the page is important. The user sees the overall layout before reading the individual steps or words. Even if the procedure is well written and clear, the user may decide against reading the procedure if the text is packed too densely on the page.

Research results demonstrate that it is easier to read and understand shorter lines of text. This is because we tend to take in a few words at a time, moving our eyes across the page in a jerking motion. In addition, a page with text laid out from margin to margin looks intimidating, especially if the lines are closely spaced. This is sometimes referred to as a "gray" page.

An open page, with shorter, adequately spaced lines, is seen by the user to be friendlier and easier to read. However, this may lead to a procedure with many pages. In an effort to save paper or to reduce the number of pages, sometimes the temptation is to use every available inch of the page.

1. VERIFY lubrication system OPERATING.
2. OPEN doors to lubrication system solid state controller.
3. CONNECT battery:
   a. LOOSEN screws on lubrication system electrical enclosure.
   b. OPEN door to electrical enclosure.
   c. PLUG battery wire leads into mother board socket.
   d. CLOSE door to unit electrical control box.
   e. TIGHTEN screws on door to unit electrical control box.
4. VERIFY all electrical connections to lubrication system are securely fastened.
5. VERIFY electrical junction box covers are CLOSED.
6. TURN ON lubrication system power switch on solid state controller.
7. While adjusting air pressure using the air pressure regulator, PRESS and HOLD MIST PRESSURE button on solid state controller keypad.
8. When header pressure is 15 inches H_2O, RELEASE PRESSURE button.

GRAY Page with text running from margin to margin and no spacing between steps

1. VERIFY lubrication system OPERATING

2. OPEN doors to lubrication system solid state controller.

3. CONNECT battery:

   a. LOOSEN screws on door to lubrication system enclosure.

   b. OPEN door to electrical enclosure.

   c. PLUG battery wire leads into mother board socket.

   d. CLOSE door to unit electrical control box.

   e. TIGHTEN screws on door to unit electrical control box.

4. VERIFY all electrical connections to lubrication system are securely fastened.

5. VERIFY electrical junction box covers are CLOSED.

6. TURN ON lubrication system power switch on solid state controller.

7. While adjusting air pressure using the air pressure regulator, PRESS and HOLD MIST PRESSURE button on solid state controller keypad.

8. When header pressure is 15 inches H_2O, RELEASE PRESSURE button.

OPEN page with shorter lines of text and spacing between steps

Figure 4-1. Example of how line spacing affects readability

Fewer pages do not necessarily result in a shorter procedure. Rather, this method results in a darker, more difficult to use document which may not even be read. These two competing criteria of document length and page darkness are in direct conflict and must be balanced. An open, easy-to-read page is more important than the desire to shorten the number of procedure pages.

Line spacing and length are often a function of the font type or size. Choosing a font type and size is a somewhat subjective human factors. As a rule of thumb, however, 12 point fonts are easily read under most lighting conditions. Anything smaller than 8 point may be hard for most users to read. An open style gives the page a professional look, makes information easier to find and read, and helps to increase the users' confidence in the procedure.

A type of page layout which effectively uses space and is easy to follow is the "T-format". The T-format divides the page into two columns which can vary in width depending on the type of information you intend to put in each. For example, as shown in Figure 4-2, the narrower left column can be used to identify the person performing the step. The wider, right column contains the actions. Notice how the wider left margin results in shorter lines which can be more easily read by the user. In the example, the procedure step is shorter since the actor is not identified in each step. This technique may be used for writing procedures when different persons or organizations (or "actors") have responsibility for the actions required to execute the procedure.

This format is also helpful when many major operations occur in the procedure. T-format helps to organize longer procedures into more easily handled modules or units. In the example shown as Figure 4-3, the major operations are identified in the left column. This helps to guide the readers to the proper step if they are interrupted or if the procedure is performed over a long period of time.

ACTOR	ACTION
Operator A	1. CLOSE transfer valve V-123.
Operator B	2. VERIFY V-123 is CLOSED on monitor 3-1.
	3. PRESS START button to start charging sequence
	4. NOTIFY Operator A that charging sequence has started.
Operator A	5. THROTTLE V-456 to maintain 120 psig on gauge 1-2.

Figure 4-2. T-format example identifying actors

OPERATION	ACTION
Preparation	1. VERIFY valve V-111 is CLOSED.
	2. OBSERVE ambient room temperature (gauge 3-3).
Charging	3. OPEN discharge V123.
	4. OPEN drum spigot.

Figure 4-3. T-format example identifying major operations

4.5.2. Hard Copy and Electronic Access Considerations

Many facilities are installing information networks to help distribute and control procedures. The primary advantage is that workers can access the most current procedures at any time for an onscreen review or can print them out for use or for training. However, electronic display has its drawbacks. Often, a procedure formatted for hard copy use does not provide a fully visible page on the monitor screen. If procedures are to be used primarily by someone on a network and are not printed out, the format should be designed specifically for the monitor display. Some other drawbacks of electronic displays are noted below:

- It is difficult for more than one or two persons to view a monitor at the same time. A printer could be valuable if a work team is going to review a document together before use.
- Observe yourself as you read a document. Users tend to flip back and forth between pages when reviewing a hard-copy document, comparing previous information, or looking for a specific step. This may not be easy when reviewing a procedure on a monitor screen.
- Computers fail. Without regular backup and uninterruptible power supplies, a completely electronic system can fail, leaving you high and dry.
- The computer workstation may not be located at the actual worksite.

4.6. Introductory Sections

Using a consistent procedure format ensures that the completed procedure will direct the reader in an unambiguous, easy-to-follow manner. In addition to the title and other identifying information, a typical procedure format might include sections that contain the following types of information:

- **PURPOSE**—The goal or intent of the procedure
- **REFERENCES**—Any documents that support development or use of the procedure
- **PRECAUTIONS**—Conditions, practices, or procedures that must be observed to avoid potential hazards involving products, chemicals, or equipment which may cause loss of life, severe injury, or which may adversely affect site operations
- **SPECIAL TOOLS/EQUIPMENT**—Any special operating, maintenance, or protective equipment or materials that must be in place, available, or obtained before starting the procedure
- **PREREQUISITES**—Any initial conditions the worker must satisfy or actions that need to be performed before starting the procedure
- **PROCEDURE STEPS**—The steps to perform to achieve the procedure's purpose safely and efficiently

Your particular format may use different names for the procedure sections, but the same information should be addressed. Additionally, you may choose to include the following sections in your procedures:

- Definitions/Acronyms
- Responsibilities
- Table of Contents (reserved for lengthy procedures)
- Acceptance Criteria (testing and inspection procedures may need this)
- Safety Systems

Whatever section titles you choose to include in your procedures, every section should be included in each procedure. Even if a section is not needed for a given procedure, leave the section title on the page, followed by *None Applicable*. This tells the user that a section has not been accidentally omitted and is an important human factors consideration. Inconsistent formatting may confuse the user and undermine the procedure's effectiveness. The sections selected for your procedures should be detailed in your procedure management system.

Figure 4-4 shows the Purpose, References, Precautions, Special Tools/ Equipment, Prerequisites, and procedure steps as they may appear in a procedure. Refer to Appendix F, *Sample Procedure Formats* (page 130), for more examples.

4.6.1. Purpose Section

When preparing a procedure, you must first determine why you are writing it and what goal or intent is to be met. State the procedure's goal in the

CCPS CHEMICAL CO.	MARYLAND FACILITY	OP-004 Rev. 001
	Selecting a Procedure Format	
PURPOSE	This example demonstrates a sample procedure format.	
REFERENCES	Sample Procedure Formats in Appendix F.	
PRECAUTIONS	This is only a sample. Procedure format should be customized to meet your facility's needs.	
SPECIAL TOOLS/ EQUIPMENT	Word processor	
PREREQUISITES	Procedure management system has been designed, reviewed, approved, and implemented.	
PROCEDURE	1. READ this book. **CAUTION** This is only a sample. Procedure format should be customized to meet your facility's needs. 2. SELECT the format concepts that meet your needs.	
	END	
April 1996.	OP-004 Rev. 001	Page 1 of 1
Approval Date: April 15, 1996.	Required Review Date: April 15, 1997	

Figure 4-4. Sample procedure layout

Purpose section. The Purpose section provides a guidepost to direct the writer (and reader) through the remainder of the procedure. The purpose statement may include the job position(s) responsible for performing the activity. The main points to remember when writing the purpose statement are conciseness and accuracy.

The following are some examples of the purpose of various procedures

PURPOSE This procedure explains how to start the XYZ reactor and bring the process to steady production after a solvent wash.

PURPOSE This procedure describes Operator, Control Person, and Technician activities to remove air and water from the unit before loading hydrocarbons for unit startup from cold shutdown.

PURPOSE This procedure describes steps for the Level 4 Technician to shut down the XYZ reactor during a loss of cooling emergency.

PURPOSE This procedure describes steps to replace a defective seal on super-pressure pump P-3266.

PURPOSE The procedure describes the steps to perform the preventive maintenance for the seven vessels in acid service in the tank farm.

4.6.2. Reference Section

After determining the purpose, develop a list of reference materials. The references will give you an understanding of the equipment and systems required to fulfill the procedure purpose. They may describe how the system or process interacts with other systems or processes. References may include

- Piping and Instrumentation Diagram (P&IDs)
- Process flow diagrams
- System descriptions
- Specification sheets
- System design specifications
- Government, international, company, or industry standards
- Process Hazards Analyses
- Material Safety Data Sheets (MSDS)
- Job and Task Analyses (JTA)
- Job Safety Analyses (JSA)
- Vendor and supplier manuals
- Existing procedures

The references listed in this section provide background or are needed to perform the procedure. MSDS and Process Hazards Analyses are examples of background information. Other procedures or specification sheets are references that could be needed to actually complete the procedure steps. The use of references within the body of the procedure should be minimized.

When a procedure requires the user to refer to another document to identify an operating parameter, maintenance specification, or regulatory requirement, there must be assurance that the reference is correct. If a referenced procedure is required to perform the task, it must also be readily available. For example, many facilities have created a single document that compiles all limiting conditions and consequences of deviation to achieve this for key operating and maintenance parameters. This concept works, but the referenced document must be current and available.

The other choice is to specify the data within each procedure that requires it. A potential disadvantage of this choice is that you may end up with several documents that require revision when the parameter changes. Each site must choose the method that works best.

4.6.3. *Precautions Section*

Once you identify the steps to perform the procedure, you must determine what precautions are necessary.

The Precautions section contains information necessary for the user to avoid injury, accidental chemical releases, and equipment damage. This section includes information on hazardous materials or unusual environmental hazards that may be present during the operation. Level of detail should be based on training and user knowledge. Within the Precautions section you should consider what can happen, why, and the consequences of ignoring the precaution. Beware of overusing precautions. The precautions should add to the procedure and address only important safety issues. The use of irrelevant or self-evident precautions will diminish the impact of the genuine safety concerns.

Quality control requirements also may be stated in this section. Plant conditions and system operations often dictate specific limitations on how or when to operate equipment or processes. A maintenance task may need to be reperformed if an acceptance criterion is not met.

The following are some examples of operations precautions

PRECAUTIONS Reactor temperature should not exceed 130°C because exceeding this temperature can result in overpressuring the reactor.

The main dump valve is extremely hot due to steam flow. Thermal gloves must be worn to prevent injury.

Charging chute collar seal must be in place at all times to prevent release to the atmosphere.

Turbine overspeed trip can result in loss of cooling flow to reboiler. Turbine speed must be maintained below 3500 RPM to prevent tripping.

The following are some examples of maintenance precautions

PRECAUTIONS Condensate pump operates at 200°F. Pump should be cooled to room temperature before disassembly.

Failure to torque head bolts to proper torque value may result in toluene leak due to seal failure or bolt damage.

A confined space permit is required to enter Tunnel B.

The precautions are then usually restated as a caution in the body of the procedure before the step that involves the concerns. The precautions listed in the front section of the procedure prepare the user to carryout the entire procedure. The caution listed directly before the step warns the user of the potential hazard immediately before the situation occurs. For example:

> **CAUTION**
> Reactor temperature should not exceed 130°C because exceeding this temperature can result in overpressuring the reactor.

5. Verify Reactor loop 3 operating temperature is less than 130°C on panel monitor 3-1.

4.6.4. *Special Tools/Equipment Section*

This section includes any special tools or equipment that the procedure user must have in order to perform the procedure. This prevents the procedure users from completing half of a procedure and then finding out that they need a tool to finish. The section can include personal protective equipment such as respirators, tools such as a micrometer, or objects such as spool pieces or chemicals.

SPECIAL TOOLS/EQUIPMENT	Airline respirator
	Non-sparking 3" wrench
	150 ft-lb torque wrench
	Pressure filter

4.6.5. *Prerequisites Section*

You must determine what conditions need to be met or actions performed before starting a procedure. The prerequisites could include:

- Initial alignment of components and systems involved in the procedure
- Plant status
- A listing of required support systems and personnel notifications or approvals
- Personnel requirements for conducting the procedure

The following are examples of operating prerequisites:

PREREQUISITES	Batch operation 2 or 3 must be in progress and wash cycle must be complete on monitor 3-1.
	Dated and signed Q/C Approval Form 345 accompany all charging drums.
	Fire suppression system is armed.

The following are examples of maintenance prerequisites.

PREREQUISITES	Maintenance personnel performing this procedure are Craft Level 3 certified.
	Compressor has been locked out in accordance with Lock, Tag, and Try Procedure 04-33.
	Reboiler temperature is less than 50°C.

Emergency operating procedures or abnormal operating procedures may title this section with INITIATING SYMPTOMS or AUTOMATIC SYSTEM ACTIONS instead of PREREQUISITES.

4.7. Procedure Steps Section

Ask experienced mechanics or operators to describe a simple task that they perform. You will quickly see that even simple tasks can involve many steps. The Procedure Steps section requires the greatest attention to analysis by the writer and reviewers. The section should describe, when applicable:

- How to recognize when the final goal has been achieved
- How the system, equipment or process should respond
- What to do if the system or process does not respond as expected
- Functions and features of engineered safety systems (alarms, interlocks)

Unique situations may cause you to address known process problems in the procedures. These process problems are often identified in Process Hazards Analyses.

Before drafting a procedure, you should consider step numbering options, the content, writing style, placement of notes and cautions, branching, and use of lists, tables, illustrations, graphics, and pictures. These should be defined in your procedure management system.

4.7.1. Procedure Step Numbering

There are many options a writer can choose when numbering a procedure step. In the past, the trend was to number procedures in what is called legal format:

5 PROCEDURE
5.1 Subheading
5.1.1 Step 1
5.1.2 Step 2
5.1.2.1 Substep
5.1.2.2 Substep

This is a valid numbering system, but it is not friendly to the user. The main advantage is that each item has a unique number for reference. The main disadvantage is that it does not make the procedure easier to understand or to use. The extra numbers at the third or fourth level don't really give users any information that can help their performance.

We recommend simpler step numbering systems that better consider human factors.

- Name the sections within the procedure body (for example, "Stripper Startup" or "Opening the Reactor Manway"). A number or letter can be assigned if desired.
- Use alphanumeric indicators to tell the reader to expect an instruction.
  —Number the items in a list only if they are sequential actions or instructions.
  —Use bullets or dashes for lists that can be performed in any sequence. For example:

Subheading
1. Step 1
2. Step 2
   - *Substep*
   - *Substep*
   - *Substep*

4.7.2. Procedure Step Content

You should consider the following questions when writing each procedure step:

- How is the step performed? For example, is this a physical action? Does the action involve interacting with a computer terminal/monitor, an automatic controller, or simple devices (gauges, valves, micrometers)?

- Can the actions be performed as written and in the sequence written? Can the equipment be operated as specified? Can the steps be physically performed? Does the user have the training or experience to understand and carry out the action using the information available, or is additional information needed?
- Does the user need to be alerted of any potential hazards (Cautions or Warnings) or need any supporting information (Notes) before performing the action?
- Does the user need to know specific operating ranges or limits to
  —Perform this action
  —Recognize the successful completion of the action
  —Recognize an actual or potential problem
  —Make a decision

  For example, does the user have to verify temperature or pressure visually? Is the user waiting for a computer command or verification? How should the process react? Will exceeding a limit result in an accidental chemical release, personal injury, or quality concern?

  What is the chance of not operating within safe operating limits, and what would be the consequences? If operation of the component is critical and a deviation will probably cause a failure, state the corrective actions. If failure is unlikely, keep going.
- Is needed information found on an instrument, panel, or monitor or is it in the procedure or another source such as a graph, table, drawing, or specification sheet? Should this information be included in the procedure or be referenced?
- What is the next logical step? How is the next step affected by what is performed in the current step?
- What are the results of improper task performance? If an action is critical, spell it out in detail. For example, when starting a special pump that may bind or cavitate if not vented and primed properly, the procedure may require more information than simply stating "start the pump." The following example illustrates that situation:

STARTING ABC PUMP

CAUTION
Failure to prime ABC pump will result in pump impeller damage.

3. **OPEN priming valve (yellow) three turns.**
4. **When water flows from top drain, CLOSE priming valve.**
5. **PRESS ABC pump START button.**

- Is the action frequently performed? Is it an action that is easily overlooked? Is this a complex piece of critical equipment that has a low frequency of use? These questions can help determine the level of detail in your procedure.
- Is the action performed so infrequently or is it so complicated that the user is not sure how to do it? Is the action so complicated that nobody is ever certain it's done right the first time?
- Is the decision point clearly defined if a decision is required? Unclear decision points can cause arguments and delays in performing actions.

4.7.3. Writing Style, Format, and Terminology

Maintaining and using a consistent style, format, and common terminology gives the user clues about what to expect, much like the shape of road signs or the legend on a map. This is part of the human factors aspect of procedures.

Determine ahead of time how you are going to identify and name equipment, tools, switches, buttons, controllers, etc. This equipment may already be labeled in the field or in drawings. When you refer to something in a procedure, it must be clearly and consistently identified. Always identify these components the same way in all procedures. Similarly, personnel referenced in a procedure should be clearly identified.

Determine ahead of time how you will use terms such as VERIFY, ENSURE, and CHECK. For example, unless otherwise specified, the terms OPEN or CLOSE would mean completely open or closed.

Changes in format or style can confuse the reader, waste his or her time, and cause loss of confidence in the procedures.

- Keep sections in the same order within and between procedures.
- Use the same page and step numbering schemes for all procedures.
- Always identify Notes, Warnings, and Cautions the same way in all procedures.
- If you call it an XYZ, call it an XYZ whenever you refer to it.
- If you spell a component name one way the first time, always use the same spelling.
- If you CAPITALIZE or <u>underline</u>, be consistent.
- Write in the present tense, using the active voice and beginning with an imperative. ("You" is the implied actor.)

- Keep procedure steps as simple as possible, with a maximum of two related actions for each step. If a step describes more than one action, the second action should be directly related to the first and performed as a single function. For example:

  —CLOSE and LOCK cabinet.
  —UNROLL and DRAIN hose.

- Keep superfluous, "nice-to-know" information to a minimum. For example, telling an operator that the inlet feed pump is green may or may not be important. If there is only one inlet feed pump, it is not important. If there is a red and a green inlet feed pump, it is important.

One thing to be wary of is the ease with which word processors allow a writer to **make bold**, *italicize*, underline, or change ***FO*nts** and **size**. If a writer becomes too free with these capabilities, they lose their special meaning. Overuse of these features results in what is known as the "Ransom Note" effect. This is distracting to a user and does not contribute to understanding.

Not so long ago, many procedure writers thought they HAD TO WRITE IN ALL CAPITALS. They assumed they were writing something important and they wanted to make sure it wasn't neglected. Now most writers are aware of the effect all capitals have on a reader. Procedures written in all uppercase text SHOUTS at the reader resulting in text that is more difficult to read. The uppercase text creates a block effect where all of the words take on a similar shape and begin to look alike. It becomes harder to recognize instructions at a glance. Decide ahead of time how you will use all capitals and other special features of the word processor. Be selective, communicate, and stick to your rules.

4.7.4. Lists and Tables

Users find lists and tables easy to understand because of the mental checklist they promote. Maintain parallel construction when writing lists. Lists also provide an opportunity to use tables in the procedure. Many word-processing programs make it easy to create a table for grouping information into columns with headings. A column for signoffs or data collection can easily be added to a table.

The following is an example of listing items:

8. PLACE these valves in the positions indicated:

VALVE IDENTIFICATION	POSITION
1-2222-1 block	OPEN
1-3333-1 automatic	OPEN
1-4444-1 block	CLOSED
1-1111-1 isolation	OPEN

Use lists or tables to simplify presentation of information. If you have three or more items or clauses in a sentence or procedure step, list the items. This is illustrated in detail in Chapter 5, *Elements of Effective Procedures*.

If multiple actions are closely related you may list items under the step. If the actions need not be performed in sequence, list items in one of the following ways:

- Alphabetical or numerical order
- Order of importance, most critical to least critical
- Physical proximity of equipment
- Order of likelihood of occurrence

If the sequence of actions in a list is critical, list each item as a separate step, or state the importance of following the sequence. For example

8. CLOSE the following valves in the order indicated:

   a. 1-2222-1 block
   b. 1-3333-1 automatic
   c. 1-4444-1 block
   d. 1-1111-1 isolation

4.7.5. Notes, Cautions, and Warnings

Notes include additional supporting information needed before performing the action. Limit Notes to that information needed to perform the action safely, prepare the user for the unexpected, or help the user to know the step is complete. A Note should contain only explanatory information, not actions.

Use Cautions and Warnings to alert users to actions or conditions that could result in injury, equipment damage, environmental releases, or major quality problems. A Caution or Warning should contain only explanatory or alerting information, not actions. A Caution or Warning should appear directly before the step to which it refers, and should always use the same format. For example:

CAUTION
Opening the manway may release ammonia fumes. Full face splash shield and goggles must be worn.

5. PUT ON full face shield and safety goggles.

For additional examples, see Figure 5-9, *Actions and Warnings* (page 66).

The use of Cautions, Notes, and Warnings, or any other such terms, should be clearly defined within the Procedure Management System to refer to different level of concerns and consequences. The terms and any formatting aids using to identify them, such as fonts and boxes, should also be clearly defined so that they will be used consistently within the procedures. As with any rules, there are exceptions. Occasionally you may find a situation where a Note really should come after a step or you need to put an action in a Caution statement. The main point is to remember that these are exceptions. Don't get in the habit of writing lots of notes and cautions that hide actions from the users.

It is a good idea to include Cautions and Notes on the same page as the step to which they refer. Break the page early if necessary. Notes and Cautions should be written in the passive voice to further differentiate them from the action steps. The required action, however, should be in the following step not in the Note, Caution, or Warning.

4.7.6. Branching

Branching is used to send the user to another location in the same procedure or to another procedure or reference. In our road map analogy, a traveler may need to branch to a more detailed map for information about city streets or building locations. The following example illustrates branching:

8. DETERMINE sample pH.
9. EVALUATE pH test results using Procedure 52 Evaluating Wastewater pH Samples.
10. If sample pH is outside of acceptable range, RETURN TO Step 1.
11. RECORD sample pH on Transport Log 1.

Keep branching within a procedure to a minimum; it is difficult to maintain the numbering scheme when steps are added or dropped. Repeat the common steps within a procedure rather than have the user jump back and forth within it. Branching should be used to allow the user to skip procedure steps or go to a specific step because a well-defined condition has been met or exists.

The following terms can be used to direct users:

GO TO—	Sends the user forward in the procedure
RETURN TO—	Sends the user to a previous step in the procedure
REFER TO—	Sends the user to an outside reference

Because poorly written conditional statements introduce the potential for user error, the procedure writing guidelines should specify the rules for using conditional statements such as if, when, or then. The consistent use of these terms, and other conditional terms, prevents confusion.

The following example demonstrates using a question as part of a procedure to branch to another part of the procedure:

3. Is XYZ OPEN?
   YES GO TO Step 8.
   NO RETURN TO Step 1.

Minimize branching to other procedures. It is best that a user needs only one document to get the job done. If you must branch to another procedure, don't jump to a step within the second procedure. Only refer the procedure user to an entire procedure or a clearly defined section in a procedure.

The following example demonstrates how to branch to another procedure and to another step within a procedure.

6. Is reactor wall thickness within 0.2 inches of last recorded thickness on Thickness Testing Log?
   YES GO TO Step 7.
   NO REFER TO MNT 544, Vessel Deficiency Procedure.
7. COMPLETE Vessel Inspection Form 22.

The following example illustrates how it can be confusing if the user is sent to steps within another procedure.

5. STARTUP hydrogen compressor by performing Steps 4—8 and Steps 12—14 in procedure OPER 332 Starting Up Hydrogen Compressor.

4.7.7. Using Illustrations, Graphics, and Pictures

Whenever a simple drawing or graphic can help a procedure, use it. A simple line drawing can be essential to a procedure that involves manipulating a complex manifold. An exploded-view diagram can greatly assist a mechanic in disassembling and reassembling a component.

Some considerations when using graphics are the following:

- What are your graphics capabilities? Can you easily put the graphics into an electronic file, or must you "cut and paste" them into the master copy of the procedure?
- Electronic graphics require a great deal of memory on a storage disk. Ensure your network or hard drive is sized properly.
- Watch out for changes in the process. If you use part of a P&ID or process flow diagram in a procedure, you must update the procedure if there is a change.
- Place graphics near the steps to which they apply. Hiding them in the back as an appendix may cause them to be ignored.

4.8. Drafting the Procedure

With your Procedure Management System in place and your format selected, the procedures can now be drafted. The following are steps to prepare an effective procedure:

1. Define the purpose of the procedure.
2. Collect and study all applicable information and reference materials.
3. Determine the qualifications and skills of the intended user (audience).
4. When drafting a procedure, speak with and observe the workers who perform, or will likely perform, the task. Interview other knowledgeable personnel. While writing, envision the actual work environment in which the procedure is performed. For new maintenance activities or processes, use a discussion group approach.
5. Ensure the procedure is technically accurate and complete. Validate it—walk down or simulate performing the procedure in the field, shop, or lab to ensure the procedure can actually be used as written and reflects current practices.
6. After the procedure has been drafted and validated, it is ready to undergo the predefined review and approval cycle (see Chapter 7,

Procedure Control). Review the finished procedure against accepted plant criteria as defined in your Procedure Management System.

7. Review the procedures for a given process as a whole to ensure that the procedures, when combined with the required training, provide all necessary information for running the process as a whole. This will often identify any information gaps between procedures.

By following these steps and sticking to accepted procedure-writing criteria, you can be assured that the procedure will be effective. The procedure is now ready to be submitted for review and approval.

4.9. The Procedure Review and Approval Cycle

Every procedure must be reviewed and approved, as described in Chapters 5 and 7. Each writer should be knowledgeable in the process for reviewing and approving procedures at their site. As determined using the information presented in Chapter 7, technical, safety, environmental and quality review considerations should be established for review and approval.

Who approves a procedure and what that approval means are important to your procedure system. Typically, the approver puts his or her signature on the document to verify that the procedure has been drafted and reviewed according to the site's system and that the procedure is the standard for performing the task until the next revision.

4.10. Special Considerations for Maintenance Procedures

Maintenance activities at a site are usually considered "crafts," and it can be a new experience for a facility to ask its maintenance personnel to write and use written procedures. Maintenance procedures require special consideration depending upon the type of maintenance force your site maintains. If your site uses cross-trained maintenance personnel, your facility may need maintenance procedures that are written to a very high level of detail. The increased use of contract maintenance personnel at facilities presents a similar problem.

Referencing vendor manuals is a choice that maintenance managers can use to keep numbers of procedures manageable, but it implies another level of document control. If you reference vendor manuals in maintenance procedures, your site must possess these documents and ensure they are accessible, up to date, and accurate.

Vendor manuals often do not provide the application-specific cautions, warnings and level of detail that your site may need. Vendor manuals are usually written generically in terms of the process application and for generic models of equipment. Your maintenance procedures may need to augment this information in order to reflect your site needs accurately.

4.11. Batch Process Considerations

The basic principles of effective procedure writing previously presented apply equally well to batch processes, but batch processes also have specific procedure considerations. Batch equipment often is used to produce multiple products with the same components. Typical batch operations are:

- Staging
- Charging
- Monitoring
- Transferring
- Cleaning/Decontaminating
- Emergencies

For batch processes, there may be two types of documents necessary to conduct operations because often the same equipment is used in different configurations for different products. First, there are operating procedures that contain the steps and safety information for performing each task. Second, there are the "batch" or "recipe" or "process" sheets that contain operating parameters such as temperature, material amounts and sequencing. These batch sheets may change with each run although the actual operating procedures remain the same. The batch sheets may change several times a week, but the operating procedures are always applicable to the equipment. Together, batch sheets and operating procedures provide the necessary information to safely run a batch. It is important to develop, update, and review them together to ensure thoroughness.

When developing, reviewing, or revising batch sheets and operating procedures, they should be evaluated as a unit since they are interdependent. Your procedure management system should specify the type of information included in the batch sheet and in the operating procedures to ensure consistent development.

5

ELEMENTS OF EFFECTIVE PROCEDURES

5.1. Purpose

This chapter provides a starting point for developing a list of criteria for writing effective procedures to meet your facility's needs. Due to your operating, maintenance, and regulatory concerns, you may want to add facility-specific procedure criteria to this list, or delete criteria not applicable to your situation. This chapter also discusses the importance of these criteria and who will use them. A summary of these criteria is included in Appendix E, *Procedure Criteria Checklist*.[1]

5.2. Importance of Procedure Evaluation Criteria

Although a process or piece of equipment may be properly installed and designed, it may not be possible to operate because the controls are obstructed, hidden, or difficult to use. Similarly, a procedure can be technically correct and yet be completely ineffective. If the information in the procedure cannot be used or understood, it does not matter how technically correct the procedure is.

Procedure Evaluation Criteria (based on procedure writing experience, human factors and lessons learned including recognized formatting and presentation methods) will help to ensure your procedures are effectively written. There are many items used to evaluate the effectiveness of a procedure. While most procedure elements are based on human factors, some are based on technical, administrative, or regulatory factors. Technical factors can include technical skill requirements or quality information records. Administrative factors can include assigning a unique number to each procedure or requiring procedure approval signatures. Regulatory factors, often based on industry-accepted best practice, can include addressing operating limits and the consequences of deviations in procedures.

Establishing a procedure development and revision checklist serves many purposes. The checklist

- Encourages early identification of your procedure criteria, which improves procedure consistency
- Helps the procedure writers to know exactly what factors they must consider when developing and writing procedures
- Speeds up procedure development because the method for writing effective procedures has been clearly defined
- Helps to minimize multiple procedure rewrites and revisions

5.3. Who Will Use the Procedure Evaluation Criteria?

Procedure writers, procedure reviewers, and procedure evaluators all use the procedure checklist:

- Writers will use the checklist to improve and maintain procedure effectiveness
- Reviewers will use the checklist as part of their review and approval criteria to help them know exactly what they are looking for when reviewing the procedures
- Evaluators will use the checklist to determine if the criteria are being followed during procedure development and revision and if additional criteria should be included

5.4. Procedure Checklist Elements

In the remainder of this chapter, an explanation and, as appropriate, an example have been included for each procedure checklist item. You may want to modify or add to these to meet your specific needs.

1. The procedure title accurately describes the nature of the activity.

Misleading, inaccurate, or incomplete procedure titles can cause the operator or mechanic to select the wrong procedure. Although this is a concern for all operations, it is especially critical when the operator or mechanic is trying to find an emergency procedure. If your procedures are in a data management system with word-search capabilities, fully descriptive titles are essential to take advantage of this feature.

The examples shown in Figure 5-1 illustrate how for a given situation, there is a complete procedure title and an incomplete procedure title. At a given facility, there are three places that nitric acid can be loaded. Therefore, a procedure titled Loading Nitric Acid is incomplete because it does not describe the circumstances sufficiently.

COMPLETE	INCOMPLETE
Loading Nitric Acid into Tank Car B Loading Nitric Acid into Reactor Receiver Loading Nitric Acid into Storage Tank 5	Loading Nitric Acid
Installing ABC Mechanical Seals on ABC Pumps	Installing Seals
Repairing Tank Farm Centrifugal Pumps	Repairing Pumps
Conducting Reboiler Monthly PM	Conducting PM

Figure 5-1. Procedure titles. Effective titles contain enough detail to enable the user to locate a specific procedure.

2. Each page of the procedure displays the necessary procedure control information such as:

- *Facility or unit name or identifier*
- *Procedure title*
- *Procedure number*
- *Date of issue, approval date, required review date, and effective date*
- *Revision number*
- *Page number and total pages*

This information allows the operators or mechanics to determine if the procedure they have is current and to make sure they have every page. This information is usually included in the procedure header and footer (see Figure 5-2).

CCPS CHEMICAL CO.	MARYLAND FACILITY	OP-004 Rev. 001
	Increasing Boiler Feed to Normal Operating Conditions	
PURPOSE	~~~~~~~~~~	
REFERENCES	~~~~~~~~~~	
PRECAUTIONS	~~~~~~~~~~	
SPECIAL TOOLS	~~~~~~~~~~	
SAFETY SYSTEMS	~~~~~~~~~~	
PREREQUISITES	~~~~~~~~~~	
PROCEDURE	1. ~~~~~~~~~~ 2. ~~~~~~~~~~	
	END	
April 1996.	OP-004 Rev. 001	Page 1 of 1
Approval Date ________________		Date Effective ______________

Figure 5-2. Procedure control information. This example shows a format for including the procedure control information in the procedure header and footer.

3. The last page of the procedure is clearly identified.

Often, due to repeated use or transport, the pages of a procedure may become detached. Including the total number of pages on the procedure and using a defined symbol to indicate the last step of a procedure ensures that this element will be satisfied.

16. CLOSE distillation column valve (V-7).		
	END	
December 1996.	OP-004 Rev. 01	Page 3 of 3

Figure 5-3. Identifying the end of a procedure

4. All temporary procedures are clearly identified. A special designation such as T or the word temporary in the title or Purpose section may be used.

Temporary procedures are a special type of procedure designed to be used for a defined length of time or amount of operation. A temporary procedure should be clearly marked temporary to ensure that the procedure will only be used in the proper circumstances. Temporary procedures may be hand written; however, review and appropriate approval are still needed.

CCPS CHEMICAL CO.	OPS-2-001-TEMP Rev. 005
	Operating Compressor B
PURPOSE:	This procedure describes how to operate Compressor B using a temporary lubrication system. This procedure expires June 8, 1996 OR when the temporary lubrication system is taken out of service (whichever comes first).

Figure 5-4. Identifying temporary procedures. The use of the word "TEMP" in the header and the word "temporary" in the Purpose section clearly indicates that this is a temporary procedure. Its expiration date is defined.

5. *Every procedure has a unique and permanent identifier.*

When a procedure is deleted, the procedure number should also be deleted rather than reassigned to another procedure. If a procedure is reactivated, the original procedure number can be reused. A deleted procedure number could be reused after a defined time period has passed. To control procedures, the procedure should have some type of unique identifier. Reusing procedure identifiers can often cause confusion.

6. *The procedure purpose clearly describes the intent of the procedure.*

The procedure purpose describes what will be accomplished or performed by the operator or mechanic. The purpose must be clear, well-defined, and attainable. The purpose is the first section that a procedure user will read; therefore, the goal must be clear.

	CORRECT
PURPOSE:	This procedure describes how to start up the Toluene Reactor System after complete reactor system shutdown. Partial startup of the toluene reactor is described in procedure OPS 342.
	INCORRECT
PURPOSE:	This procedure describes how to startup the Toluene Reactor System.

Figure 5-5. Purpose sections. An effective Purpose section clearly states what will be accomplished (the goal) under specific conditions.

7. *If you have duplicate processes, the procedures are complete and accurate for each process.*

Often, one procedure will be used to describe operation of two or more similar or parallel units. However, this requires complicated cross-referencing and a listing of the different equipment in each unit. Additionally, two processes that are considered similar may, when closely examined, be sufficiently different to justify separate procedures. However, existing procedures can provide an excellent starting point for developing procedures for similar processes. If a decision is made to have one procedure address two units, precaution must be taken to ensure that the procedure clearly defines the units to which the procedure applies and reflects actual operation.

8. *All information necessary for performing the procedure is included in the procedure or referenced.*

Branch judiciously. References should not be used to make the writer's life easier. Sending the user to other procedures or documents may make writing the procedure easier, but it makes the procedure very cumbersome. Operators and mechanics often complain that the number of references in a procedure is so great that it makes the procedure impossible to use. For example, a startup procedure refers to a compressor startup procedure which in turn refers to a turbine startup procedure, and so on. If you are just using a step or two from another procedure, copy the step to the procedure you are writing.

When you are having the user branch to another procedure, the reference should be to the entire procedure. Having the user branch to only one or two steps in another procedure can lead to errors and confusion.

Also, references must be readily available to the procedure users when they are performing the procedure. In the example shown as Figure 5-6 on the facing page, the steps to stop a leaking pump (steps 4–7) are duplicated in another procedure or in a clearly defined section in a procedure. However, rather than have the user go to that procedure for four steps, they are included in this procedure.

9. *All items referenced in the procedure are listed in an introductory section of the procedure (section often called References).*

Listing all documents identified in the procedure steps in an introductory section serves two purposes. First, it allows the procedure user to see all of

CORRECT

1. ACKNOWLEDGE inlet pump low flow alarm.
2. Is acid pump leaking?
   YES—GO TO Step 3.
   NO— GO TO Step 8.
3. OPEN backup pump inlet valve.
4. PRESS Backup Pump Start button.
5. PRESS Local Pump Stop button.
6. CLOSE pump inlet valve IV-4.
7. EXIT procedure.
8. REFER to System A Trouble Shooting Procedure TS-101, LOW FLOW.

INCORRECT

1. ACKNOWLEDGE inlet pump low flow alarm
2. If the acid pump is leaking:
   a. PERFORM Steps 1–4 of Responding to Leaking Acid Pump procedure MNT-020.
   b. EXIT procedure
3. If pump is not leaking, TROUBLESHOOT pump.

Figure 5-6. Referencing and branching. In the correct procedure, the reader does not have to refer to a different procedure for a few instructions. Additionally, the reader is referred to a specific troubleshooting procedure.

the documents required to perform the procedure. Second, it makes reviewing and revising references in a procedure easier.

10. Items listed in the References section of the procedure are correctly and completely identified.

If references are necessary to perform the procedure, they must adequately identify the document so the procedure user can locate it. References should be identified so that any procedure user, regardless of experience level, can locate the proper document. Use widely accepted, common terms when identifying reference materials.

CORRECT

REFERENCES Nitric Acid Dilution Procedure 21-003
Nitric Acid Unloading Record Sheet (Form 1)

1. If nitric acid concentration is greater than 67 degrees Baumé, PERFORM Nitric Acid Dilution Procedure 21-003.
2. COMPLETE Nitric Acid Unloading Record Sheet (Form 1).

INCORRECT

REFERENCES Dilution procedure

1. If nitric acid concentration is greater than 67 degrees Baumé, DILUTE nitric acid using dilution procedure.
2. COMPLETE record sheet.

Figure 5-7. References. The correct example gives the name and identification number of the referenced procedure or sheet so the user can easily find the correct one. In addition, the referenced procedure is identified in the Ref-

11. Write procedures using traditional upper- and lowercase letters. Use of fully capitalized words may be acceptable in certain circumstances.

Studies have shown that people read text printed in upper- and lowercase faster and more accurately than text printed entirely in capital letters because the lowercase letters add shape to the word. The overall shape of the word and the shape of the letters both contribute to word recognition. Bolding can add emphasis for critical information. Use all uppercase letters for words that need emphasis or to match equipment or instrument labels, as shown below:

- Equipment references (R36, MOV-002)
- Readings (PRESS HI, LOW LEVEL M4)
- Special information (CAUTION, WARNING)
- Actions (OPEN, CLOSE)

You should decide ahead of time how you will use capitalization, bold, italics, and other formatting issues.

12. Cautions, Warnings, and Notes used in the procedure are placed immediately before the step to which they apply.

A Caution, Warning, or Note is most effective if it alerts the procedure user to a situation before the situation occurs. Cautions, Warnings, and Notes should be separated from the text to be easily distinguished from the steps. The format and content of these sections should always be presented in the same manner and format so that the procedure user knows where the information will be within the procedure.

Occasionally you may find a situation where a Note really should come after a step or you need to put an action in a Caution statement. The main point is to remember that these are exceptions.

CORRECT

2. PLACE controller PIC-123 in MANUAL mode.

> **CAUTION**
> When opening valve B, the operator should stand behind the steam turbine to prevent contact with steam.

3. While standing behind steam turbine, OPEN Valve B.

INCORRECT 1

3. OPEN Valve B carefully while preventing exposure to steam.

INCORRECT 2

4. OPEN Valve B.
   You may be burned by hot steam if you are standing in front of the valve when it is opened.

Figure 5-8. Cautions. In the correct example, the user is warned of the potential hazard before performing the action rather than after performing the action.

13. Cautions, Warnings, and Notes should stand out from the procedure steps.

Cautions, Warnings, and Notes should contain the explanatory information that describes the conditions and/or unusual circumstances necessary to safely perform the procedure. Additionally, Cautions and Warnings should include any relevant consequences of the potential situation. If action statements are imbedded within the textual explanations, they may be overlooked. Writing them in the passive voice helps to differentiate the explanatory information from the steps that must be performed. Even if

CORRECT

7. CLOSE reactor purge valve.

> **WARNING**
> When opening the reactor manway, there is a potential for exposure to toluene and benzene. An airline respirator must be worn.

8. PUT ON airline respirator.
9. OPEN reactor manway.

INCORRECT

7. CLOSE reactor purge valve.

> **WARNING**
> When opening the reactor manway, there is a potential for exposure to toluene and benzene. Purge the reactor and put on airline respirator.

8. OPEN reactor manway.

Figure 5-9. Actions and warnings. In the correct example, the required action is defined within a step rather than hidden within the WARNING where it could easily be overlooked.

the active voice is used within Cautions, Warnings, or Notes, the action must be repeated as a numbered step to visually prompt readers that they must perform an action.

You may decide to use icons, different sized or style font, or other attention attracting characters instead of, or in addition to, boxes to differentiate Warnings, Notes, and Cautions. Again, the idea is to make these items stand out from the procedure text. Examples of icons can include familiar road signs and transportation symbols such as a stop sign.

14. If more than one person is required to perform a procedure, the procedure is written for the person responsible for performing each step.

If a procedure requires more than one person to safely perform it, the procedure should clearly indicate who is responsible for performing each step. Industry studies have found the need for a more consistent and structured approach to these types of procedures. This allows the procedure to be easily evaluated to determine if all steps and communications are included for every person involved in the activity. Otherwise, critical steps

CORRECT	
Mechanic	1. CONTACT inside cat operator by radio (Channel 2) to verify communication link.
Inside Reactor Operator	2. VERIFY unit pressure on controller PIC-32 is less than 0.5 psi. 3. If pressure is less than 0.5 psi, NOTIFY mechanic it is safe to remove head bolts.
Mechanic	4. REMOVE Unit Head Bolts.

INCORRECT
1. CONTACT inside cat operator by radio (Channel 2) to verify communication link.
2. VERIFY unit pressure on controller PIC-32 is less than 0.5 psi.
3. If pressure is less than 0.5 psi, NOTIFY mechanic it is safe to remove head bolts.

Figure 5-10. Specifying responsibilities. In the correct example, the person responsible for performing a step is clearly identified.

may be omitted when the person performing the procedure changes or because there is no clear assignment as to who must perform a step. This example shows how different persons are identified.

15. A signoff line is provided for verifying critical steps of a procedure.

This is optional, depending on the potential hazards that may be involved in the performance of a procedure. If a step is critical, such as one step that could cause injury, accidental chemical release, degraded product, or equipment damage if performed incorrectly or omitted, a signoff line provides assurance that the step has been completed. Refer to Figure 5-11.

16. When the procedure requires coordination with others, checklist, signoff, or other method can be used to indicate that steps or actions have been performed or completed.

A checklist or other device can be used to record notification of action or completion of an action. This is critical when coordinating a multiperson procedure to prevent omission of steps, especially if the steps are critical to operations or safety. If the procedure is performed in a high stress condition, such as an emergency, this checklist/accounting device reduces the likelihood of error or omission due to stress induced error. The accounting device also provides a record for after-the-fact documentation of emergency or unusual situations.

The following example shows a checklist to ensure that actions required in the event of high pressure in a tank have been completed. These actions are performed by different people at different locations.

Action	Initials/Time Performed or Completed
1. Inlet Valve V-123 CLOSED	__________
2. Inlet Valve V-456 CLOSED	__________
3. Tank T-34 DRAINED and VENTED	__________
4. Emergency Fan EF-001 RUNNING	__________

Figure 5-11. Checklists. In this example, a space is provided for the person performing the task to indicate who completed the task and when it was completed.

17. Procedure steps are written in short, concise sentences.

Studies show that short, concisely written steps increase the ease of reading and comprehending the instructions. "Nice-to-know" information should be minimized. The ideal step contains an action verb and the object of the action.

For example, STOP (action verb) the pump (object of the action). As necessary to enhance understanding or performance, the following may be included:

- Action limits — OPEN valve 30%.
- Object identifiers — OPEN pump discharge valve.
- Location — OPEN pump discharge bypass valve located downstream of expansion joint.

CORRECT

13. MONITOR cooling water flow rate on LI-26.

    NOTE: Valve B4 is located behind control panel.

14. Using valve B4, MAINTAIN cooling water flow rate between 20 and 25 gpm.
15. NOTIFY supervisor if flow rate drops below 20 gpm.

INCORRECT

13. MONITOR cooling water flow rate to ensure water flow rate is not too high. Cooling water flow rate is adjusted using valve B4. If cooling water flow rate falls, supervisor must be notified.

Figure 5-12. Using concise steps. Effective procedures contain concise, unambiguous language.

18. If a step contains more than two items, they are listed rather than buried in the text.

When a step contains many items, the use of a list or table makes it easier to find and track the items. It also helps to prevent the user from missing or overlooking an item.

CORRECT

12. OPEN the following compressor valves:
    - Bypass valve V-634
    - Discharge valve V-876
    - Inlet valve V-234
13. Visually INSPECT the following relay contacts for pitting:
    - Intermediate relay
    - Starter relay
    - Time Delay relay

INCORRECT

12. OPEN compressor inlet valve, discharge valve, and bypass valve.
13. Visually INSPECT the intermediate relay, starter relay, and time delay relay for contactor pitting.

Figure 5-13. Listing items in procedure steps. In the correct example, items are listed for easy reading, not hidden in a long string of text.

19. If two actions are so closely related that they are performed as a single action, a single step may make more sense.

Sometimes actions are so closely related that it does not make sense to separate them into two steps.

CORRECT

9. LOOSEN and REMOVE four bearing cap bolts.
10. REMOVE upper bearing cap.
11. REMOVE bearing from upper cap.
12. ROTATE shaft counterclockwise to roll bearing out of bottom housing.
13. PLACE both bearings in cleaning solution.

INCORRECT

9. LOOSEN and REMOVE four bearing cap bolts
10. REMOVE bearing cap, then ROTATE shaft in a counterclockwise manner to loosen bottom bearing, then REMOVE bearings and PLACE both bearings in cleaning solution.

Figure 5-14. Related actions. In the correct example, more than one action is placed in a step only if the two actions are closely related or performed as a single action.

20. Procedure steps that must be performed in a fixed sequence are identified as such.

The term procedure implies a sequence. The steps in a procedure *should* be performed sequentially. In some situations it may be critical that the steps be performed in a certain order to prevent hazards or equipment damage. In other situations, order may not be an important issue.

CORRECT EXAMPLE 1

23. TIGHTEN head bolts in the following sequence: ◇ = Bolt

Front

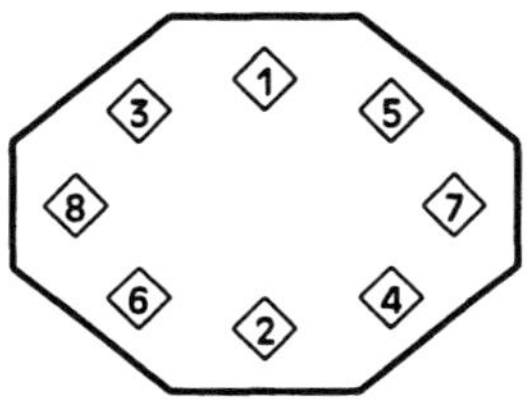

Back

CORRECT EXAMPLE 2

CAUTION
Failure to install relays in correct order may result in faulty operation.

24. INSTALL relays in the following order starting from port 18.

	Port	*Relay*
a.	18	R-1
b.	19	R-3
c.	20	R-2

Figure 5-15. Emphasizing fixed sequences. This example shows the use of a diagram and a Note to emphasize that the steps must be performed in a specific order.

21. Operating or maintenance limits or specifications are written in quantitative terms.

When performing a procedure, the operator or mechanic should have all information necessary to perform that procedure, including limits of operation. These limits should be presented in clear, quantitative terms. The limits should not be described qualitatively, because that leaves room for interpretation. Plus or minus limits should not be used since these require mental calculations.

EASY TO USE	HARD TO USE
300 psig (255 psig to 345 psig)	300 psig ±15%
Increase temperature 5°C per minute	Increase temperature quickly
50 ft-lbs (45 to 55 ft-lbs)	50 ft-lbs ±5
From 5.00 to 5.25 inches	Large enough so the shaft does not bind
Fill beaker to the 100-ml mark	Fill beaker ½ full

Figure 5-16. Quantitative terms. The easy-to-use examples give the user the necessary information without requiring mental calculations which may waste time or lead to error.

Additionally, procedure steps should use units as they are actually read or measured by the user. For example, if a tank level gauge is measured in inches, the step should not refer to percentage full.

22. Procedures should provide instructions for all reasonable contingencies. If contingency instructions are used, the contingency statement precedes the action statement.

Many procedures are written as if operations or maintenance tasks are constant and that exceptions to the standard method for performing a task do not occur.

CORRECT 1

> **CAUTION**
> Forcing shaft can damage coupling.

25. INSERT shaft ¾ inches.
26. If shaft binds and cannot be fully inserted, PLACE shaft in freezer for 10 minutes.

CORRECT 2

23. START drain pump (DP-01).
24. MONITOR drain pump sight gauge.
25. If tank cannot be drained to 0 inches on sight gauge:
    a. OPEN valve EV-001 to process drain sump.
    b. NOTIFY Waste Treatment to expect high pH load.
26. STOP drain pump (DP-01).

Figure 5-17. Identifying contingencies. Examples 1 and 2 show the steps for handling contingencies.

The instructions for contingencies should be included in the procedure to inform those who cannot rely on their experience, such as new or inexperienced personnel, to correctly and safely respond to the situation. The use of conditional terms, such as *if*, *when*, and *then*, should be predefined to ensure that all procedure writers use the terms in the same way.

23. If conditions or criteria are used to help the user make a decision or recognize a condition, the conditions precede the action.

A criterion or condition is a type of prerequisite. Therefore, the condition should be stated before the action that is dependent upon it. Otherwise, a procedure user may not read the step completely and may perform the action without satisfying the requirement.

CORRECT
1. VERIFY plant air system is operational.
2. START hydrogen flow to reactor at 5 cfm.
3. MONITOR distillation column outlet temperature (TI-42).
4. If distillation column outlet temperature (TI-42) is greater than 45°C, SET feed flow to the distillation column at 18–35 gpm.
INCORRECT
1. START hydrogen flow to reactor at 5 cfm after plant air system is operational.
2. MONITOR distillation column outlet temperature (TI-42).
3. SET feed flow to distillation column to 18–35 gpm if distillation column outlet temperature (TI-42) is greater than 45°C.

Figure 5-18. Identifying conditions or criteria. If conditions must be satisfied to perform a step, the condition is stated before the step to prevent performing the step and then reading the required condition.

24. Conditional instructions should be easy to understand. If more than one condition requires the same sequence of actions, list each condition separately.

Often, before an action is performed, several requirements must be satisfied. These conditions should be presented so they can be easily read and understood. If the conditions are embedded in the step, the procedure user may miss one or more of the conditions. Errors can occur if a procedure user overlooks, misinterprets, or misunderstands a contingency because the contingency is not clear. These errors are even more likely to occur in high stress situations.

CORRECT

9. If any of the following conditions exist:
   Reactor temperature is greater than 200°C (TI-33)
   —OR—
   Reactor pressure is greater than 150 psig (PI-124)
   —OR—
   Reactor feed flow rate is less than 80 gpm (FI-22)
   a. OPEN reactor inlet valve.
   b. PLACE cooling water switch in REFRIGERATED.
   c. ADD 40 gallons of water to reactor.

INCORRECT

9. If reactor temperature is greater than 200°C (TI-33), or reactor pressure is greater than 150 psig (PI-124), or reactor feed flow rate is less than 80 gpm (FI-22):
   a. OPEN reactor inlet valve.
   b. PLACE cooling water switch in REFRIGERATED.
   c. ADD 40 gallons of water to reactor.

Figure 5-19. Conditional instructions. In the correct example, the conditions are listed for ease of reading rather than hidden in a string of text.

25. When procedures require calculations, the calculations must be clear and understandable. In order to ensure accuracy for complicated or critical calculations, a formula or table is included in the procedure or on a referenced calculation sheet.

Having the calculation in the procedure or readily available ensures accuracy and consistency. Retaining the calculation provides a record of the results and ensures the accuracy of the calculation. If the procedure users perform the calculation on scrap paper or even worse in their head, errors can occur. Additionally, unless auditors can access all of the data, such as the results of calculations, they will have difficulty determining what happened when unexpected results are obtained.

For complex or critical calculations, a means should be provided for recording the result.

CORRECT

1. CALCULATE the total weight of sulfuric acid to be loaded.

   Weight H_2SO_4 = 15 pounds/gallon H_2SO_4 × (Volume H_2SO_4 in gal.)
   Total weight H_2SO_4 = ____________

2. Record results on Tank Truck Loading Form (Form 12-94).

INCORRECT

1. CALCULATE the total weight of sulfuric acid to be loaded into the truck.
2. Record results on Tank Truck Loading Form (Form 12-94).

Figure 5-20. Calculations. The correct example includes the calculation and all other necessary information to perform the calculation.

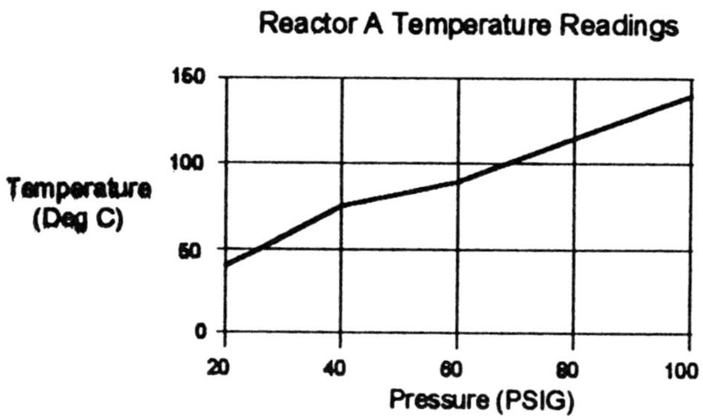

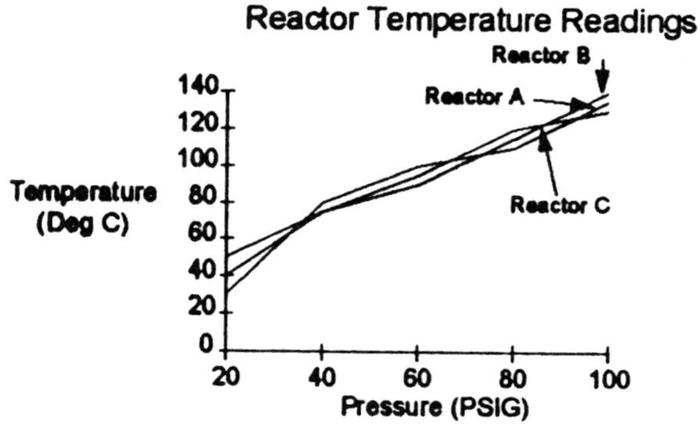

Figure 5-21. Graphs. In this example, the user is only responsible for Reactor A. The top graph can be easily read and only contains information about Reactor A. The bottom graph is cluttered and difficult to read.

26. Graphs, charts, and tables in procedures are designed so that values can be easily and accurately extracted and interpreted. Minimize the information in tables and graphs to present only the information necessary to perform the required action.

Graphs and tables are an excellent tool for providing important information to the procedure user. Similar to developing a concise, clear procedure step, graphs and tables should be presented in a way that ensures that the information can be easily understood and accessed. (See Figure 5-21.)

27. Procedures that specify alignment such as valve positions, pipe and spool configurations, or hose station hook-ups should
- ***Individually specify each item (do not refer to other steps)***
- ***Identify each item with a unique number or designator***
- ***Specify the position in which the item is to be placed***
- ***Indicate where the user records the position***

Systems are often misaligned because the procedure lacked detail or was confusing.

CORRECT

4. CLOSE the following valves:
   a. Inlet feed pump discharge valve (V-201).
   b. Inlet feed pump inlet valve (V-737).
   c. Acid storage tank discharge valve (V-126).
5. PLACE Standby Feed Pump (P-27) switch in AUTOMATIC.
6. PLACE Feed Pump switch (PS-27) in OFF.
7. ALIGN feed flow to the standby inlet feed pump (P-39):
   a. OPEN standby feed pump discharge valve (V-916).
   b. OPEN standby feed pump inlet valve (V-917).
   c. OPEN acid storage tank discharge valve (V-918).

INCORRECT

4. CLOSE inlet feed pump discharge valve, inlet feed pump inlet valve, and acid storage tank discharge valve.
5. PLACE standby Feed Pump switch in AUTOMATIC.
6. PLACE Feed Pump switch in OFF.

Figure 5-22. Valve alignments. In the correct example, the user has no doubt as to which pieces of equipment should be operated and what must be done.

28. An Emergency Operating Procedure contains provisions for verifying

- *Conditions associated with an emergency*
- *Automatic actions associated with an emergency*
- *Performance of critical actions*

Emergency procedures are almost always performed under high stress conditions. Therefore, the procedure must be presented in a way that guides the user through the emergency, including notifying the user of symptoms associated with a potential emergency situation.

A Operator

1. If Material Release Alarm sounds:
   a. ACKNOWLEDGE and SILENCE Release alarm.
   b. Visually INSPECT sump level.
2. If sump level is below 10 feet:
   a. STOP #1 Sump Pump.
   b. CLOSE emergency drain valve (EDV-003).
   c. RECORD alarm on SUMP LOG 1.
3. If sump level is between 10 to 15 feet:
   a. OPEN emergency drain valve (EDV-003).
   b. START # 1 Sump Pump.
   c. MONITOR sump level.
4. If sump level exceeds 15 feet or third level alarm sounds:
   a. NOTIFY Manager to initiate Emergency Response Plan.
   b. LEAVE area.

Figure 5-23. Emergency Operating Procedure. In an effective emergency operating procedure, the conditions associated with the emergency and the necessary actions are clearly defined.

29. The procedure includes any required followup actions or tests and tells the user who must be notified. This typically occurs when a procedure involves turning over a system from operations to maintenance or vice versa.

Sometimes tests, alignments, or other actions must be performed before a piece of equipment or a system can be placed in or returned to service. In other cases, if a component fails a test, actions may have to be taken to replace or discard that item. This is especially critical for procedures that require hand-off to another party, for example operations to maintenance or maintenance to operations.

CORRECT
6. ALIGN agitator pump drive motor in accordance with Vendor Manual VM-0093.
7. NOTIFY electrician to make electrical connections.
8. VERIFY electrician has completed connections.
9. VERIFY inspection sticker is: • Dated • Signed • Affixed to motor casing.
10. REMOVE personal lock and tag in accordance with plant lockout procedure (SAF-020).
INCORRECT
6. ALIGN agitator pump drive motor in accordance with Vendor Manual VM-0093.
7. NOTIFY electrician to make electrical connections.
8. REMOVE personal lock and tag in accordance with plant lockout procedure (SAF-020).

Figure 5-24. Followup action. In the correct procedure, the user's followup actions after electrical work are detailed in steps 8 and 9.

30. If a procedure must be performed by someone with a special qualification, the procedure must specifically state the required technical skill level to perform the task.

This ensures that the worker performing the task has been qualified to perform the task. The qualification requirements may be included in the Prerequisites or Purpose section at the top of the procedure. The following are some examples of qualifications

- Mechanic must be a certified welder.
- Mechanic must have Level 7 training and qualification.
- Mechanic must be Level C craftsman.
- Operator must be qualified in use of airline type III respirators.
- Only Type 3 lab technician may sample #2 day tank.

Endnotes

1. "Procedure Criteria Checklist" and the elements listed in this chapter were adapted from NUS Training Corporation's *Procedure Writing Workshop Manual*. Gaithersburg, MD, Seventh Edition, 1995.

6

WRITING EMERGENCY OPERATING PROCEDURES

6.1. Purpose

This chapter discusses several factors to consider when identifying and writing procedures for abnormal or emergency conditions. We define an operating emergency as a condition caused by an event which can result in an accidental chemical release or injury if appropriate actions are not taken. Such actions are described in Emergency Operating Procedures (EOP). These actions may range from stabilizing the process for continued operation to shutting down the process in a controlled manner.

Emergency Response Procedures, as defined in your Emergency Response Plan, for major on- or off-site releases, fires, or explosions are not considered in this book. However, the interaction of EOPs with Emergency Response Procedures is discussed.

6.2. Defining Events Requiring Emergency Operating Procedures

EVENT An identifiable occurrence related to equipment performance or human action, or an occurrence external to the system that causes (or could cause) dangerous system upset. The cause or contributor to an incident or accident.

SYSTEM UPSET A condition where any system operating parameter falls outside established safe operating limits, resulting, or likely to result, in unstable operation, operation outside of design limits, or potential release.

In this chapter we will emphasize system upsets or abnormal situations where the loss or degradation of components or controls could allow process parameters to exceed the design intent or limit of the process or equipment, resulting in an accidental chemical release. The time factor involved in an operator's ability to assess and correct an operating deviation is discussed in more detail in CCPS' *Guidelines for Integrating Human Factors into Process Safety Management Systems*.[1]

The cause of a system upset may be understood. For example, it may be caused by the loss of instrumentation or utilities such as cooling water. Or the cause may be unknown or not fully understood; for example, a runaway reaction, or a high-temperature or high-pressure situation. In either case, predefined procedures must be followed for continued safe operation or shutdown.

An operating emergency could also result from the failure of a vessel, pipe, valve, or other component to contain or control a hazardous chemical. The ability of an operator to safely and quickly respond to an operating emergency helps protect the people, the environment, property, and plant equipment.

Operating emergencies can also result from human error; for example, introducing the wrong chemical into a system or not following a procedure correctly may result in an operating emergency.

An example of an event requiring an Emergency Operating Procedure would be a high-pressure, high-temperature, or high-level situation where the failure to perform the correct actions in a timely manner could result in an accidental release. This is especially critical when the process is designed to run close to the safe operating or design limits. The loss or degradation of the following items may lead to system upsets:

- Utilities (cooling, heating, electric power, ventilation, inerting/blanketing gases, process and breathing air, hydraulics, natural gas, fire water, telephone, etc.)
- Controls
  —valves
  —flow controllers/dampers
  —computer and microprocessor control
- Instrumentation
  —sensors and indicators (electronic, pneumatic; local, remote)
  —transmitters/transponders
  —meters/gauges
  —alarms and annunciators
- Equipment (vessels, pipes, pumps, agitators, compressors, etc.)
- Redundant or backup safety systems (mechanical or electronic)

Emergency Operating Procedures may also be written for situations where there may be advance notification or warning, including natural occurrences such as tornadoes, hurricanes, blizzards, floods, or man-made occurrences such as bomb threats. Depending on the seriousness of the situation and time limitations, proper responses may include accelerating safe process shutdown or placing the process in a safe or self-sustaining condition.

Written emergency operating procedures are very important and effective tools. The user must understand the critical steps needed to regain control or stabilize the process to prevent an incident. Appropriate actions may prevent a process upset from escalating into an emergency.

6.3. Identifying Emergency Situations

Potential emergency situations may initially be identified by a properly performed Process Hazards Analysis (PHA). Other sources of potential events include incident reports as well as user or plant experience with similar processes or equipment at your plant or other plants. Various CCPS Guidelines contain additional information on process hazards analysis[2] and quantitative risk analysis.[3]

After potential events have been identified, they must be prioritized. Typical risk ranking methods consider the probability or frequency of occurrence and the criticality or potential for injury or environmental impact. You should direct your initial efforts to developing those EOPs that address the greatest risk.

Alarms do not necessarily indicate an emergency. Alarms only alert an operator or technician to a plant condition. The plant conditions determine if an emergency exists. The conditions causing the alarm may still be within safe operating limits for the process, but may be approaching conditions that would qualify as a system upset. A spill or overflow condition may not require an EOP if a containment system is in place to contain and control spills and prevent a safety or environmental impact. In some cases, however, an alarm may initiate an EOP. For example, a high-temperature alarm may initiate emergency actions requiring operating personnel to isolate the cause of the alarm, to shut down the process, or to immediately evacuate the area or facility as directed by the facility's Emergency Response Plan.

6.4. Developing and Writing Emergency Operating Procedures

Effective Emergency Operating Procedures can keep events from escalating by instructing the operator to

- Perform critical actions
- Make effective decisions quickly
- Focus attention on the emergency

Operating procedures should specify a clear "entry" into an EOP. The EOPs should identify a clear "exit" to operating procedures. Operating procedures may include steps or sections which send the user to an EOP if certain defined conditions exist. Likewise, conditions that may require invoking the Emergency Response Plan may be included in the Emergency Operating Procedures. (This point is discussed in detail in Section 6.8.)

Effective procedures can help the operator determine when an operating emergency exists by defining initiating conditions. An Emergency Operating Procedure should describe the best actions to take to mitigate or stabilize the situation. An EOP may simply direct the operator to press an emergency stop button and then evacuate to a safe area, or it may direct the operator to reroute process flow. EOPs will instruct the operator to take action. This could include shutting down the affected process units. The following conditions could require initiating an EOP:

- Failure of components (pipes, pumps, vessels) to contain or control hazardous materials
- Critical operating information about the unit (pressure, flow, temperature, level, composition, etc.) is unavailable due to loss of instrumentation
- Safety systems, including backup, are inoperative
- Other systems or equipment which support normal operations or normal shutdown are not available (Example: feed, product, or cooling medium flow may be lost because of equipment failure or because pipes are plugged or blocked)
- Valves normally used to control the process cannot be used
- There is an emergency such as a fire or release in another part of the facility
- Other conditions exist—for example, severe weather conditions or bomb threats

In all cases, the EOPs must be easily accessible, concise, and to the point.

Emergency Operating Procedures may require the user to start emergency backup equipment or support systems. These systems may start automatically, requiring operator verification and monitoring, or they may have to be manually activated or physically connected to the process and manually operated. Systems normally under automatic control may have to be operated manually. Auxiliary equipment or infrequently operated components may have to be put on-line to control the situation. Loss of systems or instruments can compound the emergency. EOPs should indicate actions to take for such failures. For example, the loss or unavailability of a cooling water system may require the operator to use another cooling source. The failure of a system could even require immediate evacuation and initiation of the facility's Emergency Response Plan.

Some questions you should ask yourself before writing EOPs include:

HOW will the Emergency Operating Procedure be used?

- From memory—training used to ensure consistent response
- Quick reference—memory jogger
- Step-by-step instructions—perform the steps in sequence
- Status verification—a checklist used to verify that the proper actions were taken

WHERE are Emergency Operating Procedures to be used?

- At the equipment
- In the main control room
- In an auxiliary control room

WHO is going to use the Emergency Operating Procedures?

- Control Room Operator
- Equipment/Process Operator
- Supervisor/Manager
- Mechanic
- Other trained personnel

The same standards and methods used for developing, writing, and controlling operating and maintenance procedures (Chapters 3, 4, and 7) are used for EOPs. You must remember that EOPs are intended for situations where timely actions are of the utmost importance. EOPs must serve as job aids, not as job hindrances. Precious time will be lost if a user must search for the right procedure or wade through pages of explanation before finding the critical information needed to control the situation. Procedure titles should be clear so the users are not confused or uncertain about having the correct procedure. In many cases, the use of simplified drawings, tables, or flow charts may be as effective as written steps. Emergency operating procedures are written in the same way as other procedures.

The following is a list of some types of information you might include in an Emergency Operating Procedure:

- Acknowledging and silencing alarms
- Responsibilities for performing specific actions during the emergency
- Appropriate personal protective equipment (PPE) or other protective devices (showers, eye wash stations, first aid supplies, extinguishers, fire monitors, deluge cannons, emergency carts)
- Special tools, materials, or chemicals
- Additional hazards not present during normal operations
- Location and use of emergency equipment
- Location of alternate control stations or panels
- Location of manual stops and shutoffs for systems normally under automatic control
- Decision aids (flowcharts, drawings)
- Safe operating limits and other indicators
- Shutdown lists, diagrams
- Consequences of operating outside safe operating limits
- Steps to place the process in a safe or self-sustained mode
- Steps to shut down the involved process or threatened processes in the safest, most direct fashion
- Instructions detailing conditions when by-passing emergency shutdown systems or interlocks is allowed
- Steps to return the process to safe operating limits if possible or practical
- Steps to maintain control within safe operating limits
- Conditions under which the user may have to stop and evacuate
- Required communication, announcements, and notifications, including initiating the Emergency Response Plan

When considering the above list, remember: the time, equipment, or systems used for controlling the process in the normal manner may not be available. Loss of equipment and controls may mean the user cannot cool, purge, or drain the process before or after the emergency.

The actions described in an Emergency Operating Procedure may be based on Process Hazards Analyses, drills, simulations, or previous experiences at your own or similar facilities. If there is more than one possible response, depending on the cause or nature of the emergency, you should first list instructions with the greatest likelihood for success, considering level of loss and availability of plant resources (backup systems and personnel). For example, the most likely cause for a "bad" stack emission could be an economizer tube leak. The least likely cause may be a product cooler water leak. The EOP should be written to direct the user first to the economizer and the steps required to isolate it, then to the cooler if the economizer is not the cause of the "bad" stack.

Unlike operating or maintenance procedures, Emergency Operating Procedures cannot always be easily tested. It is not always possible to address all potential conditions that may cause an emergency. However, if you base the EOPs on sound information from Process Hazards Analyses and other good engineering, operating, and maintenance practices, you enhance the effectiveness of the EOPs.

Training plays an essential role in ensuring appropriate response to emergencies. Operators should be trained in the purpose, use, and limitations of EOPs. Training in how to handle upset conditions can speed up response time and lessen the adverse effects of such events.

6.5. Directing the User to the Correct Emergency Operating Procedure

The user must be able to locate the appropriate EOP to respond to an emergency situation easily and quickly. Users should not be forced to look through procedures or procedure manuals to find the proper response. This would require them to remember which procedure contains the appropriate EOP and where in the procedure or manual the EOP is located.

One method to ensure quick access to the appropriate Emergency Operating Procedure is to include the EOP condition and title under a unique heading at the end of an operating procedure. Because users are trained that instructions and references are always at the end of the procedures, they can quickly go there for the appropriate information. This section may reference other EOPs or may contain specific instruction steps. If you include EOP instructions in one procedure, you should standardize the location of the EOP instructions in every procedure.

If a response is unique to a specific procedure or piece of equipment, you may want to include the instructions in the procedure (for example, leak in Reaction Vessel X is handled differently than any other reaction vessel) or you may find it more practical to write a stand-alone procedure. In either case, the operating procedure would direct the user to the required information.

OPERATING PROCEDURE

REFER TO the identified procedure for listed condition:

Condition	EOP Reference
Fire Alarm	Emergency Response Plan
High XYZ Monitor Alarm	EOP 011 Responding to High XYZ Monitor Alarm
Plate Cooler Leak (low cooler discharge pressure <5 psig)	EOP 001 Responding to Plate Cooler Leaks
Reaction Vessel Line Leak	EOP 020 Reaction Vessel Line Leaks

Figure 6-1. Using a table to identify common EOPs

1. If pressure falls below 5 psig or a leak develops in aftercooler AC1 eflux line, REFER TO EOP-203 Aftercooler Reflux Line leak.

Figure 6-2. Example of an operating procedure step explicitly stating the conditions which trigger an EOP

1. After all tanks have been emptied in accordance with this procedure, CONTINUE with normal shutdown in accordance with NSD-003 Normal Shut Down of Aftercooler System.

Figure 6-3. Example of an EOP step returning the reader to an operating procedure after conditions have been stabilized

When an EOP describes a condition common to many components or areas (for example, a general pipe leak), the operating procedure may reference a single "common" EOP. This minimizes the chance that the response to the same type of incident, for example, a plate cooler leak, could differ from one procedure to another. The common EOPs should be written for all applicable equipment. Any referenced procedures should be identified in the procedure's reference section to facilitate training and reviews.

The example in Figure 6-4 gives the user clues for assessing the seriousness of the situation. It incorporates the techniques for writing effective procedures.

REACTION VESSEL 1X FEED LINE LEAK

Control Room Operator

1. Are any fumes visible?

   YES— REMAIN in control room.

   a. NOTIFY maintenance of situation (ext. 1234). Maintenance will notify ERT.

   b. CLOSE automatic valve AV1.

   c. WAIT for arrival of ERT for instructions and permission to leave.

   NO— GO TO Step 2.

2. After verifying fumes are NOT visible, ISOLATE leak:

   a. (Inlet Line 1) CLOSE valve IV2 and valve IV3.

   b. (Inlet Line 2) CLOSE valve IV4 and valve IV5.

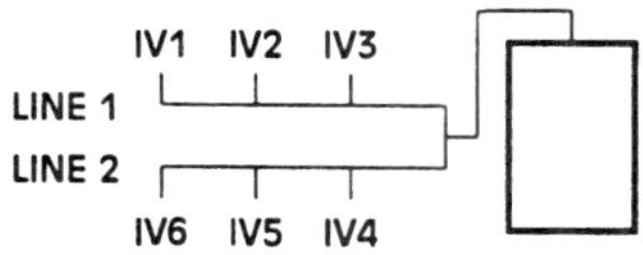

3. If fumes become visible, RETURN to control room and CALL Maintenance (ext. 1234).

4. WAIT for arrival of ERT for instructions and permission to leave.

Figure 6-4. Using visual indicators to help the user assess the situation

6.6. Incorporating Human Factors in Emergency Operating Procedures

Keep in mind that Emergency Operating Procedures are written for use in the "heat of battle." The user may be, and usually is, under great stress. Communications systems may be compromised or lost. Knowledge of the status of other processes may be limited or nonexistent. Power may be out or available only to some equipment or areas; lighting may be poor. Confusion may be rampant, but immediate response is required; there is no time to read a procedure for understanding. The EOP must help the user to make quick, safe, efficient decisions and to perform actions with high probabilities for success. Effective EOPs can also act as a confidence booster. At the same time, the EOPs must help the user access and use available information so that actions are based on actual conditions.

As with operating and maintenance procedures, you want the user to use the written EOP to control the situation. Theerefore, EOPs must be accessible and must address the specific needs of the user. From a human factors standpoint, how the EOPs are accessed can affect their use. Here are a few methods to make the EOPs easy to locate and use during an emergency:

- Place the most critical procedures in the front of the procedure manual
- Ensure manuals and tabs are clearly labeled
- Use colored paper for EOPs
- Use colored tabs to separate the EOPs from others procedure types
- Place the EOPs in a separate manual, with a brightly colored cover and spine
- For electronic access, identify a "hot key" that is always active to bring up the EOPs

Training and drills help reinforce the habit of using the EOP. During some emergencies a user may have to perform immediate actions from memory. The EOP should be obtained and used as soon as possible.

6.6.1. Ease of Reading

The user must be able to read the procedure easily. The following are some techniques that may be used to make Emergency Operating Procedures easy to read:

- **Use LARGER than usual type**
- Use simple, direct steps
  1. CLOSE Valve A.
  2. PRESS E Stop located above control panel.
  3. REPORT to assigned assembly area.
- Use simplified drawings showing the location of manual shutoff or isolation points
- Use lists
- Use lists in conjunction with simplified drawings (see Figure 6-5)

6.6.2. Giving Clear Directions

Instructions should direct the user to the necessary information for assessing and controlling the situation. This may be a trend graph on a computer terminal or a current condition readout on an indicator. Sometimes the process data may no longer be available from the automatic control system,

EMPTYING TANK A TO RX1

1. CLOSE following valves:
   —VC
   —VE
2. VERIFY the following valves are OPEN:
   —VA
   —VB
3. VERIFY arrow on 3-way valve VF points to pump P1.
4. START pump P1.

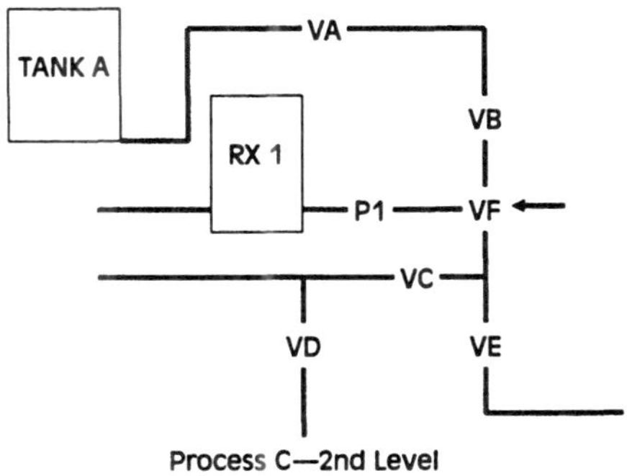

Figure 6-5. Example of simplified steps with a very simple drawing

but is available only from a local indicator on a remote control panel. As much as is practical, include parameters and information needed to understand the situation in the EOP. The use of graphs and tables can be the easiest way to present this information. In the following example, the acceptable pressure range for PI-46 is 50% of the acceptable range for PI-123. The example includes the actual pressure ranges rather than percentages so that the user does not have to perform calculations.

TAKING PRESSURE READINGS TO CONTROL R6 DURING VISIBLE FUMES

NOTE: PI-46 is located behind east pipe chase. The normal pressure on PI-123 should be between 140 and 200 psig.

1. If pressure indicator PI-123 is not accessible because of excess fumes, READ gauge PI-46.

NOTE: Pressure reading on PI-46 should be between 70 and 100 psig.

2. If tank pressure on PI-46 is outside of acceptable range, OPEN valve V3.

Figure 6-6. Example of an EOP directing an operator to instrumentation which is used only when the normally used indicator is inaccessible.

6.6.3. Availability

Other human factors questions which should be asked include:

First, how are accurate and up-to-date Emergency Operating Procedures going to be made available to the user? There are many methods for providing information to the user, but all methods must be controlled.

- **Electronically**—Procedures may be available from a terminal or printed on a local printer. In this case consideration must be given to possible power failure and the need for appropriate backup. In addition, the terminal should be located where it can be most conveniently viewed during an emergency, and users must be able to access the document easily.
- **Preprinted Copies**—These are the procedures you are probably most familiar with. These EOPs must be readily accessible to the user. They must be maintained current. (See Chapter 7, *Document Control*.)
- **Placards or Posted Written Procedures**—These can be located on or near components and emergency shutoffs or stops (see Figure 6-7).

IN CASE OF FIRE
This alarm is for EVACUATION ONLY.
It DOES NOT notify the Fire Department.

1. SOUND this fire alarm:
   BREAK glass and PULL lever
2. If safe to do so, TELEPHONE Fire Department at 3333.
3. LEAVE by nearest exit.
4. If unable to do so earlier, TELEPHONE Fire Department at 3333.

Figure 6-7. Example of a posted, written procedure.

Again, as with any procedure, electronic or hard copy, these documents must be controlled to ensure that the information is accurate and up to date.

- **Quick Reference Cards**—Quick reference cards may include wallet or pocket size cards that include information such as alarms, phone numbers, assembly points, main shutoffs, or emergency actions. These cards should be easy to read and should address specific conditions and actions. As with any procedure, the cards must be controlled to ensure that the information is accurate and up to date.

FRONT

PLANT ALARMS
CONTINUOUS SIREN— **PLANT EMERGENCY** GO TO assigned evacuation point outside fence ER Team ASSEMBLE
MODULATING WAIL—FIRE Area 1 only: GO TO designated assembly point All other areas: WAIT for instructions Fire Team ASSEMBLE
INTERMITTENT SIRENS— **AREA EMERGENCY** ALL Personnel: LEAVE AREA GO TO designated point ER Team ASSEMBLE
ALL CLEAR Announced over Plant PA

BACK

ASSEMBLY POINTS
CALL EXT. 9999 if unable to go to assigned area.
AREA 1 EVACUATION—South visitors lot FIRE—North contractor lot AREA EMERGENCY—Warehouse 3
AREA 2 EVACUATION —South contractor's lot FIRE—as instructed over plant PA AREA EMERGENCY—Fire House
AREA 3 EVACUATION—South administrative Lot FIRE—as instructed over plant PA AREA EMERGENCY—Admin. Bldg. Lunchroom

Figure 6-8. Example of a quick reference card

If placards, posted instructions, quick reference cards, or booklets are used, the instructions must not conflict with other applicable written procedures. These instructions must be controlled in the same manner as written procedures.

Second, where are manuals, terminals, or printers located?

- Will they be in or near areas where they will most likely be needed and away from potential exposure to fumes, spray, or smoke? Or, will they be on the side of a tank or piece of equipment or in a location where smoke, fumes, or spray may make it difficult, if not impossible, to see and read the instructions?

6.6.4. Physical Indicators

In addition to alarms, printouts, and instrument indications, the human senses are valuable indicators of a potential emergency. These should also be included as appropriate. These are

- **Sight:** Smoke, fuming, flame, sparks, color, spills, position
- **Hearing:** Strange sounds or lack of sound
- **Smell and Taste:** Acrid or sweet odors or taste sensations
- **Touch:** Unusual heat or vibration

> 1. While monitoring for visible fumes, if you have a metallic taste in your mouth:
>   a. LEAVE area.
>   b. NOTIFY Supervisor.

Figure 6-9. Example of human senses used as an indicator of potential emergency

6.7. Using Decision Aids

There are many ways to help the user quickly make decisions based on available information. Simple, direct methods make the user's task easier.

Some of these techniques include the use of

- Specific questions which can be answered yes or no to quickly move user to applicable steps or procedures
- Flow Charts and Decision Trees
- Decision Tables

The use of specific questions which are answered yes or no is effective. Avoid requiring users to read through many "if type" questions before finding the right step or information.

The example shown as Figure 6-10 includes steps that can be quickly answered yes or no to address a system cooling alarm. This allows the user to quickly find the appropriate instructions for the actual conditions.

Decision tables display information used to make emergency operating decisions in an easily accessed format. They are also sometimes referred to as contingency tables.

The example in Figure 6-11 identifies the problem, then quickly directs the user to the appropriate step in the procedure.

Flow charts and decision trees are symbolic representations that show the progression of operations in an activity. These charts can include decision points that direct the user to other actions based on a simple yes or no response (see Figure 6-12).

RESPONDING TO REACTOR 7 HIGH-TEMPERATURE ALARM

1. IS system pressure greater than 100 psig?
   YES— GO TO Step 4.
   NO— GO TO Step 2.
2. IS temperature greater than 100°C?
   YES— GO TO Step 4.
   NO— GO TO Step 3.
3. IS cooling water pressure greater than 25 psig?
   YES— GO TO Step 4.
   NO— GO TO Step 6.
4. Is auxiliary cooling pump 3 operating?
   YES— NOTIFY supervisor and EXIT procedure.
   NO— START auxiliary cooling pump 3.
5. WAIT five minutes, then RETURN TO Step 1.
6. ACKNOWLEDGE alarm.
7. INCREASE cooling monitoring frequency to once every 30 minutes.

Figure 6-10. Example of the use of questions in a procedure

Indication: EXCESS STACK EMISSIONS (white vapor from West Stack)		
COMPONENT	IF	GO TO
Economizer	Economizer Gas Exit TEMP less than 85°C (TG-01)	Step 4
Product Cooler	Product Cooler Exit Temperature greater than 120°C (TG-02)	Step 8

Figure 6-11. An example of a decision table that could appear in an EOP or Quick Reference Card

6.8. How Emergency Operating Procedures Link to the Emergency Response Plan

Your plant's Emergency Operating Procedures and Emergency Response Plan are two entirely independent documents and are intended to address different types of emergencies. However, as mentioned in Section 6.4, an EOP may identify conditions that, if satisfied, would require that an operator initiate the Emergency Response Plan directly or by notifying another party. As applicable, EOPs should indicate when and how the

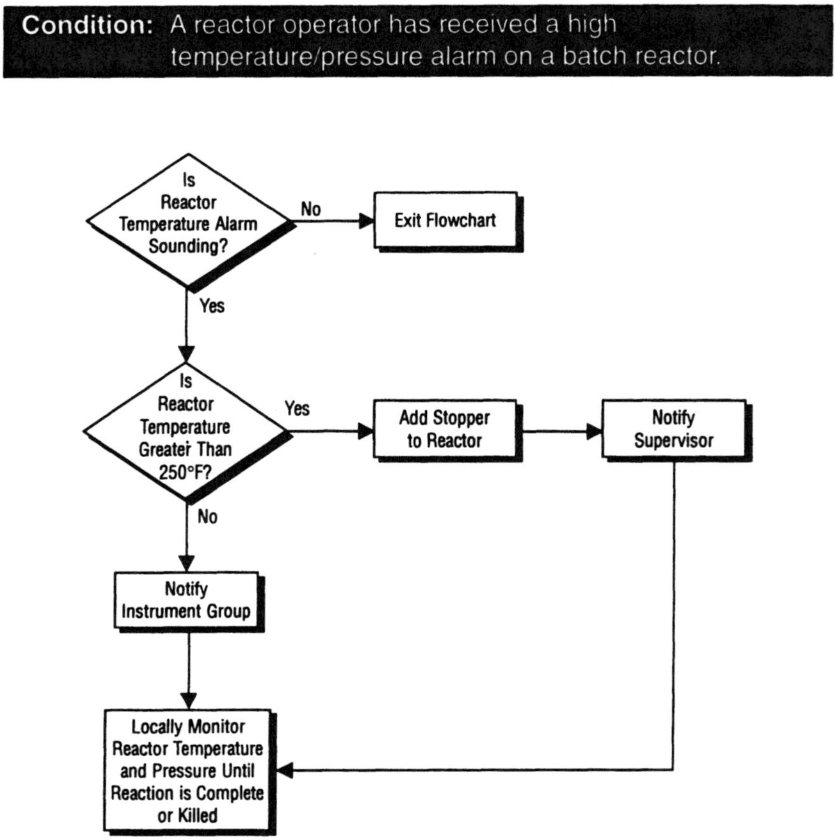

Figure 6-12. Sample Emergency Operating Flow Chart

Emergency Response Plan should be invoked. This may involve sounding an area alarm or notifying a preassigned party when certain conditions are apparent. Additional actions may include sounding an alarm, shutting down the process, and evacuating. The decision to evacuate or initiate the Emergency Response Plan is unique to each site. Personnel who can invoke the Emergency Response Plan should be identified.

EOPs should indicate the person responsible for performing the emergency operations and, when necessary, initiating the Emergency Response Plan. If the EOP is written for use by a single user, this may be evident. If more than one person is involved, the procedure should assign responsibility to specific people or positions. The example in Figure 6-13 assigns specific responsibilities to the Operator, Supervisor, and Manager. It also gives limits and actions for initiating the Emergency Response Plan.

RESPONDING TO PRODUCT RELEASE ALARM

Operator

1. OPEN Emergency Product Dump valve.
2. MONITOR for airborne levels.
3. If airborne levels exceed 2,000 ppm or if the Release Alarm sounds:
   a. CLOSE all inlet valves to reaction vessel RC1.
   b. NOTIFY Supervisor.
   c. EVACUATE to assigned assembly point.

Supervisor

4. ACKNOWLEDGE and SILENCE Release alarm.
5. VERIFY Release alarm panel lights and area beacons are LIT.
6. NOTIFY A and B control rooms.
7. MONITOR airborne levels.
8. If airborne levels exceed 5,000 ppm or third level alarm sounds, NOTIFY Manager to initiate Emergency Response Plan.

Manager

9. Based on input and assessment of Supervisor, ACTIVATE ER siren and ANNOUNCE emergency over Plant PA to initiate Emergency Response Plan.

Figure 6-13. Example of an Emergency Operating Procedure which assigns specific responsibilities

Endnotes

1. For additional information on this topic, refer to Center for Chemical Process Safety (CCPS) publication *Guidelines for Integrating Human Factors into Process Safety Management Systems*, New York, New York, 1995.
2. For additional information on this topic, refer to Center for Chemical Process Safety (CCPS) publication *Guidelines for Hazard Evaluation Procedures*, New York, New York, 1992
3. For additional information on this topic, refer to Center for Chemical Process Safety (CCPS) publication *Guidelines for Chemical Process Quantitative Risk Assessment*, New York, New York, 1989.

7

PROCEDURE CONTROL

7.1. Purpose

This chapter presents the basic requirements to access, review, approve, control, and maintain procedures. It also discusses the wide range of choices you have in selecting the level of control your plant needs and electronic document control methods that may be of assistance. Controlling the development, revision, and implementation of your procedures is an important part of your procedure management system (described in Chapter 3, *How to Design an Operating and Maintenance Procedure Management System*).

The need to keep procedures up-to-date and accurate is a theme that runs throughout this book. Chapter 2 clearly shows that document control is a key element of process safety, environmental responsibility, and quality. A workable Process Safety Management system requires that you control the access, review, approval, revision, and maintenance of procedures. If your plant has a Total Quality Management System in place, such as ISO 9000, document control for operating and maintenance procedures is essential to keep your certification current.

7.2. What Is Procedure Control?

Procedure control can be defined as methods to

- Monitor the review and approval cycle during procedure development
- Ensure that only current versions of approved procedures are accessible to the users

- Ensure that any changes to the procedures or new procedure development are made according to set rules defined in your procedure management system
- Plan and conduct a systematic, periodic review of all procedures to ensure accuracy and to meet regulatory and facility requirements
- Ensure procedures interface with the Management of Change system at your facility

A simplified flow chart showing typical procedure control during development is shown in Figure 7-1. As shown, procedure development, revision, approval, training, implementation, and evaluation are key steps. The need for a procedure can be identified by a user, supervisor, or management during periodic review, or at any time while the procedure is in use or as a result of Management of Change, Process Hazards Analyses, change in technology, new process, change of materials, etc.

Procedure control is one component of the procedure management system described in Chapter 3. Typically, the written description of the procedure management system includes the steps for controlling procedures.

Although every facility may have different quantities of procedures, different formats, or different review and approval needs, a high-quality Process Safety Management system requires current operating and maintenance procedures. Ensuring that procedures are written and revised when needed (especially before new equipment or methods of operation are implemented) and made available to the users should be a major goal of your facility's document control system. However, achieving this purpose is not as easy as it seems. If you are just starting to write procedures, develop and implement a procedure management system first. If you already have procedures in place without a formal procedure management system, establish one.

7.3. Controlling Procedure Revisions and Development

Your procedure management system should define how an employee can request a change to an existing procedure or the development of a new procedure. Procedure changes should be evaluated in accordance with your MOC system and conducted in accordance with your procedure management system. If your facility has only a few employees, an employee may request a change verbally by notifying the nearest manager. It is a good idea to keep a written procedure request log, using a standard form. This

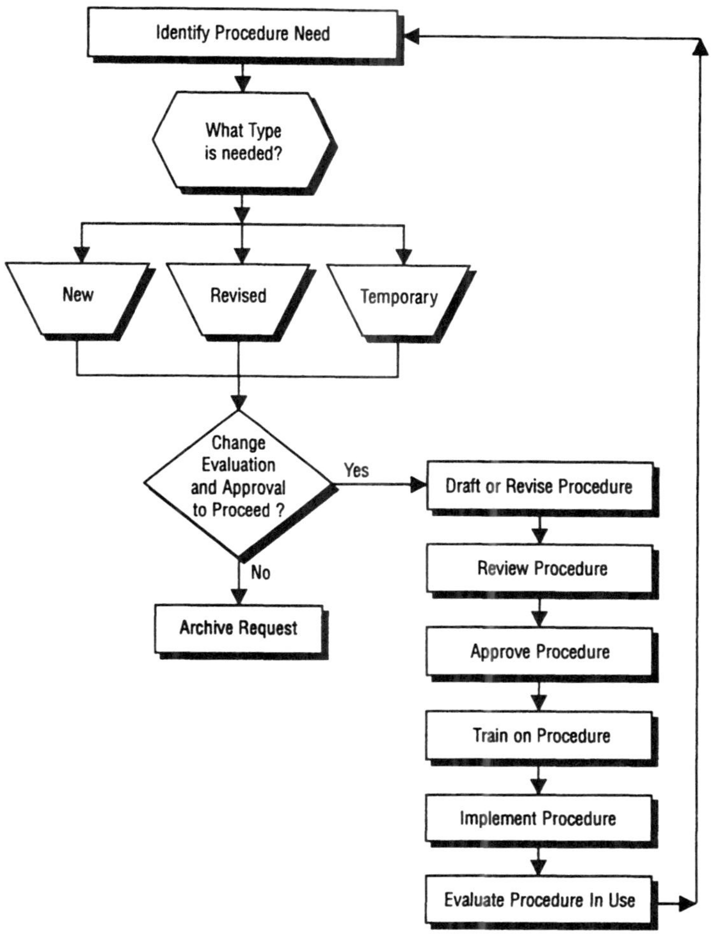

Figure 7-1. Simplified procedure control flow chart

provides a history of the requests that have been made and allows tracking of these requests.

An example of a standard form for requesting a procedure change is shown in Figure 7-2. The titles shown on the form identify functions that can be performed by existing personnel. The procedure requestor is the person who requests the procedure development or revision. The procedure sponsor is the person who is responsible for ensuring that the procedure is accurate and up to date. The procedure coordinator is the person responsible for processing the procedure change requests.

CCPS CHEMICAL COMPANY			NEW YORK FACILITY
PROCEDURE REQUEST FORM			
DIRECTIONS: Any employee may identify the need for a new procedure or change to an approved procedure. Determine the type of change needed (new–revised–temporary–deleted) and check the appropriate box in Part 1. If this is a new procedure, recommend a title and the corresponding number will be assigned. If this is a revision, write the full title and number of the current, approved procedure. Briefly describe the new procedure or change in Part 2 and note if this is a temporary procedure. Describe the reason for this request in Part 3. Complete the requestor information in Part 4. Transmit the completed form to your Procedure Coordinator for approval to proceed. (Marked-up procedures, drawings, or other information may be attached and should be listed in Part 2.)			
PART 1			
New Procedure:	Revision:	Temporary:	Deletion:
Procedure Number:		Current Revision Number:	
Title:			
PART 2			
Description of the New Procedure / Revision:			
PART 3			
Reason for the Procedure / Revision / Deletion:			
PART 4			
Requestor:			Date:
Procedure Sponsor:			Date:
Approved:		Disapproved:	
Scheduled for implementation on this date:			
If request disapproved, briefly describe why:			
Revision 0-Page 1of 1.			Approved February 1996

Figure 7-2. Sample Procedure Request Form

Adapt this form to your needs or create a form of your own. Design your procedure request form as simply as possible, making sure it reflects your management and staffing capabilities. Procedure users and other employees should be trained in the use of the form, because they are the best source for identifying procedure needs that could otherwise be missed.

If you have followed the advice in earlier chapters, you involved the users in both the development and the review phases of the procedures in their area. Users should notify the proper person at your facility if they encounter a change that needs to be made in a procedure. If a problem arises, a user can have the needed change made quickly to ensure procedures are accurate.

After identifying the need for a revised or new procedure, the procedure should be drafted in accordance with your procedure management system. This will ensure consistency and increase the efficiency of development. Organization, formatting, and basic administrative considerations are described in detail in Chapters 4, 5, and 6.

7.4. Who Should Review the Procedures?

At a small facility, procedure review may not present a problem. "We just let Julie and Jeff have a shot at it!" is a typical response. Julie and Jeff may test the procedure, mark up the draft, and send it back for approval and publishing. That was an easy example: just review one procedure at a time, and get it through the system. However, large facilities or small, multiproduct facilities writing lots of procedures face different problems. There may be dozens of procedures in different phases of development and revision at any time. How do you control this?

The first thing to do is set the criteria for procedure reviews. Determine the following:

- Who should review each type of procedure?
- What constitutes a thorough review?
- How long should it take to get a procedure reviewed?
- How do we document that a review was performed?
- How often will we review procedures?

Workers skilled in doing the job are the first logical resource for a procedure review team. Operators review operating procedures. Maintenance mechanics review maintenance procedures. Laboratory technicians and chemists review laboratory procedures. This promotes employee participation and ownership in the procedure development process.

Although you must define what constitutes a thorough review for your facility, some items that should be evaluated for each procedure are

- Safety—both process and personnel
- Technical accuracy—sound operating/maintenance practices

- Environmental concerns
- Quality—both product quality and procedure quality
- Regulatory requirements
- Accepted work practices and rules

Procedures for some jobs may not need more than a review by the users and the supervisor or facilitator. Others may greatly benefit from, or require, review by engineering, safety, environmental or quality groups at your facility. Typically, you need to specify minimum review requirements in your procedure management system, but additional reviews may be requested.

In order to avoid a lengthy wait during the procedure review process, a standard review period should be specified. If you are responsible for implementing a given number of procedures by a certain date, you need some way to get reviewers to do their part on time. One way to do this is to use a document review form. This form fills two needs: first, it lets you establish a reasonable review period (typically 5, 10 or 15 workdays); and second, it gives you a way to document each reviewer's concerns for eventual resolution. A typical review form is shown in Figure 7-3. This form records data such as when a document was sent, to whom it was sent, why it was sent, and when it should return. Specific comments can be added or attached, and you can use the form to document employee participation in the procedure process.

It is essential that the reviewers understand your facility's procedure program and procedure management system, as well as what their role is as reviewers. Procedure reviewers benefit from receiving the same training as the writers. The result is faster, more thorough reviews and higher quality procedures.

7.5. Procedure Approval

Once the procedure has been through the review cycle and the comments are incorporated or resolved, you are now ready for the approval phase. It is essential that your procedure management system specify who approves a procedure. The responsibility for a given type of procedure should be assigned to the person most directly involved with the day-to-day process activities. Typically, this person is responsible for ensuring that procedure training has been completed and that only current copies of the procedure are in use.

CCPS CHEMICAL COMPANY — NEW YORK FACILITY

PROCEDURE REVIEW FORM

DIRECTIONS: This form is used to log review comments for new or temporary procedures and procedure revisions. The writer completes Part 1 and sends the package to the reviewers. The reviewer(s) will list comments in the left-hand column of Part 2 and route the package to the writer. The writer will note how comments were resolved in the right-hand column and file the package until approval. Additional comment pages, drawings, and marked-up procedures may be attached.

PART 1

TO:____________________ (Reviewers (List All)) DATE DUE:____________________ (5 Workdays after Routing)

DATE ROUTED:____________________ RETURN TO:____________________ (Writer or Procedure Sponsor)

Procedure Number	*Rev. No.*	*Procedure Title*	
Writer:	*Reason for Submittal (check appropriate type):*		
	Revision:	New Procedure:	Temporary:

Brief Description of the Purpose (or reason for Revision)

PART 2

Reviewer Comments:	*Writer Resolution:*

Reviewer Signature ________ Date ________ Writer Signature ________ Date ________

PRE Rev.0 January 1996

Figure 7-3. Sample Procedure Review Form

The person responsible for approving the procedure should receive the same opportunity as a reviewer to comment on the document and have the writer resolve the comments before it is signed. When it is decided that the procedure is satisfactory, the procedure can be signed as approved. The approver's signature can be on a master copy of the procedure or kept in a file on an approval sheet. This depends upon your particular procedure management system.

What does it mean when a procedure is approved? It simply means that the document has undergone a systematic review against the most current available information and that cognizant persons have evaluated it against their experience. It does not mean the procedure is perfect. Procedures will always be in a state of continuous improvement. Approval of a procedure simply means it is current, accurate, and ready for use.

7.6. Evaluating Procedures in Use

Every procedure program should plan for a periodic review of the procedures. Your procedure management system should also define who is responsible for conducting the review. This evaluation will determine if the Management of Change system and the procedure management system are working properly and being followed. The periodic review involves sending existing procedures through the procedure review process. A regular review period is common for many facilities, but the time period depends on each facility's needs. The Occupational Safety and Health Administration's (OSHA's) 29 CFR 1910.119 *Process Safety Management of Highly Hazardous Chemicals; Explosives and Blasting Agents, Final Rule 1992* requires certifying the procedures are current and accurate on a yearly basis.

Two difficult considerations you will face are

- How do I make sure new procedures and latest procedure revisions are available to the users?
- How do I ensure obsolete procedures are removed from use?

The most current version of approved procedures must be made available to the users. This presents a new problem that your procedure management system can solve if it specifies the location and number of procedure sets available. If properly designed, your procedure management system will ensure that only approved procedures are available to employees, and that approved revisions replace out-of-date procedures. How you control access to procedures depends on your the delivery methods, the criteria that you set up in the procedure management system, and how your facility manages change.

7.7. Electronic Document Control

Though not essential to maintaining your procedures, an electronic document management system can assist you in keeping your procedures current and in making them readily accessible. If the system is easy to use and networked into all facility areas, it can be the simplest way to ensure that the set of procedures is kept together and updated. If the users have access to a computer and printer, they can print out a current procedure and take it into the field for use. You will not have to worry about a user removing a master copy for reproduction and forgetting to replace it.

There are some features that you may want to consider if your facility is building or buying an electronic document management system

- Networking capability and security. The system should be able to provide access to all users in all parts of the facility, both for viewing and printing, but still keep anyone from making an unauthorized change.
- Document access. The ability to organize procedures in logical categories and subcategories is essential for locating files.
- Document search capability. Operating and maintenance procedures are a vital part of management of change, and often a plant change can affect a large number of procedures. Document search capabilities allow you to find the procedures or key words within the procedures. For example, a centrifugal pump has been replaced by a reciprocating pump. You need to know which procedures include references to the old pump. Document search capabilities allow you to identify those procedures that require revision.
- Multiphase development tracking. To manage a large procedure effort, you may want to record the writers, reviewers, sponsors and subject matter experts who participate in the development of each procedure. You may want to track the progress from draft to approval. Certain electronic document management systems will help you do this.
- Periodic review scheduling. An electronic document management system can act as a tickler system to assist in scheduling periodic procedure reviews. An electronic database system can automatically start a clock each time a procedure is approved and provide a monthly or weekly report on which procedures require review.
- Automatic indexing and transmittal. Once your hard-copy locations for procedure sets are identified, an electronic document management system should help you to know where procedure sets are located and who should receive new or revised procedures.

Will electronic data management solve all your problems? No, you will still need to maintain hard copy sets of procedures. As long as there are power outages and failed computers, hard copies should be maintained. An electronic document management system will, at the least, help you keep your hard-copy sets current and intact.

8

PROCEDURE DEVELOPMENT COSTS AND BENEFITS

8.1. Purpose

This chapter summarizes why written operating and maintenance procedures are a critical part of Process Safety Management, the typical costs associated with procedure development, and the return on investment.

8.2. Reasons for Procedure Development

Written procedures are an important part of Process Safety Management because they

- Help reduce accidents
- Promote safe, efficient operation and maintenance
- Improve quality, continuity, profitability, and cost control
- Build upon and record process experience
- Help maintain your competitive edge
- Reflect best thinking on how to operate and manage your facility safely
- Comply with governmental regulations or industrial initiatives requiring written procedures

The importance of procedures is well documented and emphasized by various consensus groups and government regulations. For example, OSHA investigations conclude that poorly conceived operational or maintenance procedures are among the underlying factors that contribute to or heighten the consequences of an accident.[1]

Additionally, Appendix C of OSHA's PSM Rule states that, "An effective process safety management program requires a systematic approach to evaluating the whole process. Using this approach the process design, process technology, *operational and maintenance activities and procedures, nonroutine activities and procedures*, emergency preparedness plans and procedures, training programs, and other elements that impact the process are all considered in the evaluation."[2]

Unfortunately, according to Industrial Safety and Hygiene News' Process Safety Management Report Card, the written operating procedure requirement is the element of the OSHA PSM Rule for which most citations are written. "Many facilities don't document the procedure or they make changes and don't go through and document them formally in writing."[3]

The cost of not developing procedures is also well documented. Industry data compiled by the American Insurance Association[4] indicate that approximately one-fourth of all major accidents resulting from operating failures during a 17 year period could be tied to

- Absence of recommended procedures for sections of the plant
- Inadequate startup and shutdown procedures
- Poor training programs

Another report, Marsh & McLennan's "Large Property Damage Losses in the Hydrocarbon—Chemical Industries—a Thirty Year Review," estimated that there was approximately $7.35 billion worth of property damage during that 30-year period. That estimate does not include the cost of business interruption, extra expenses, fines, penalties, employee injuries, or liability claims. Of that amount, approximately $51.8 million were attributed to operational errors that led directly to the losses.[5]

In addition to effective procedures simply being good practice, this information alone justifies the cost of developing effective procedures for your facility. This book is intended to help you optimize your efforts and minimize your costs when developing effective procedures.

8.3. Procedure Development Costs

The only real cost is not investing in procedures.
—John McIntosh III of Procter and Gamble

The EPA's "Risk Management Programs for Chemical Accidental Release Prevention; Proposed Rule 40 CFR Part 68" estimates the cost for developing standard operating procedures at $2,500 for simple processes to

$14,000 for complex processes. However, industry experience has indicated that it can cost from $2,500 to millions of dollars to produce effective procedures for a facility.

There are many items that can affect this estimate including:

- Availability of technical information
- Accuracy of technical information
- Availability of personnel experienced with the process
- Experience and expertise of your procedure writers
- Cost for all personnel involved
- Amount of time the writers can devote to the procedure effort
- Selection of your procedure delivery system
- Design of your procedure management system
- Complexity of your process

The following sections outline the costs you can expect to encounter during a procedure development project. Three main costs associated with procedure development are the time involved to

- Implement a procedure management system
  —Design or revise system
  —Develop system and document system
  —Train procedure writers, reviewers, and approvers on the system
  —Activate system
- Draft or revise a procedure, including:
  —Train procedure writers
  —Gather information
  —Interview subject matter experts and observe task performance
  —Write the procedure
  —Validate the procedure by observing task performance or simulation
- Implement the procedure including:
  —Review the procedure and incorporate necessary edits
  —Approve the procedure
  —Train procedure users on the procedure
  —Distribute procedure
  —Conduct periodic procedure review and revision incorporation

8.3.1. Procedure Management System Design, Development, and Implementation Costs

During the design phase of developing your procedure management system, gathering relevant information, interviewing all necessary personnel, and designing the system are the costs to consider. There are also the additional review and approval costs which, depending on the complexity of your procedure management system design, can be extensive.

Since time is a major cost factor in procedure development, tracking procedure development progress is important so that you remain on schedule. There are many alternatives to consider in deciding how to track a procedure project. For the writers, a measurement of the number of first drafts sent to review is an obvious choice. The reviewers can often cause a major delay, so there needs to be some way to know where a specific procedure is at all times (and a phone number to call to get it moving). However, the only datapoint that really matters is: Are the procedures approved and available when you need them?

When you budget for your project, you will start with a list of procedure titles and a total number of procedures that can be used as a starting point for assigning writers, allotting estimated development times, and assigning reviewers. But what typically happens as you begin writing and reviewing the documents? Remember, we are organizing information so that it is easy to comprehend and helps the employees do their job. Your list of initial procedure titles will fluctuate over the course of the development process. Writers and reviewers may suggest combining some tasks or splitting others apart. As total knowledge of the equipment grows, some tasks that were missed in the initial scoping step will be added. These changes must be tracked in order to stay on top of your site's total procedure program and make schedule adjustments.

As discussed in Chapter 7, *Procedure Control*, there are software packages that can assist in project tracking and in managing future changes to these documents. Trying to keep up with a large-scale procedure-writing effort with a pencil and graph paper can be a full-time job. Procedure management software with a well-designed database for tracking drafting, reviews, comment resolution, approval, and revision history can greatly help you manage the project.

8.3.2. Procedure Drafting or Revision Costs

When writing procedures, the following time-related costs can be incurred:

- Training procedure writers in effective procedure writing
- Gathering all technical information
- Interviewing and observing the subject matter experts perform the task
- Writing the procedure
- Validating procedure content to ensure procedure reflects actual practices

The following are some questions to consider:

- Have you clearly defined your procedure criteria?
- Who will write your procedures?
- Do they have procedure writing and technical experience?
- Can they give the required dedicated time to writing the procedures?
- Are they familiar with your procedure delivery system?
- Do they have the appropriate training?

Due to deadlines or limited resources, many facilities have found it cost-effective to use outside resources to assist in writing their procedures and associated training documents. By contracting experienced procedure developers, a large number of procedures can be revised or drafted and placed into the review cycle rapidly. A well-designed project using contracted procedure developers can still offer the benefit of employee participation by using operators, technicians, process engineers, and managers as subject matter experts.

8.3.3. Procedure Implementation Costs

The costs associated with implementing the procedures may not be immediately apparent. However, implementing the procedure requires time for review, approval, and distribution—*and time is money*. These costs should be included when calculating the project budget.

Who are your procedure reviewers and approvers? Procedure reviewers and approvers must have specific training and skills as defined in your procedure management system. Even with this internal guideline, you still must put thought into selecting the specific individuals who will review and evaluate the documents.

Ask yourself the following questions:

- Do the operators, mechanics, technicians, and engineers who will serve as reviewers have enough time available?
- Does everyone on shift need to see each procedure?
- Should we plan an Engineering review for every document or just for particular ones?
- Should an Environmental, Health, and Safety representative review each procedure?

Remember, each pass through a review cycle has a hidden cost and extends the duration of the project.

One method for improving the review cycle efficiency is to train the reviewers on the site procedure management system, procedure format, allotted review time, and specific goals of the project at hand. Let them know what each person's goal is in terms of expertise. For example, operators and mechanics should look at the procedures with an eye toward past experience, human factors, equipment terminology, and step sequencing. Process Design and Engineering should focus on step sequencing, control schemes, safe operating limits, and safety systems. Safety reviews should look at worker protection and to the potential for exposure. A standard time period should be set for all reviewers to complete their part of the cycle. Train the reviewers on the procedure management system and help them focus their reviews and they will spend less time on each document but accomplish their goal.

Depending on your procedure delivery system, you may have to create hard copies of the procedures to distribute or add files to a computer system[6] to ensure that the procedure users can access the procedures. Time spent distributing the new or revised procedure and removing the old ones is another cost factor.

8.4. Return on Investment: Improvements You Can Expect from Effective Procedures

The benefits of effective operating and maintenance procedures are often hidden or difficult to isolate, but they do exist and are extremely valuable. At many facilities other efforts for improvement are in place besides procedure development—design changes, safety-related upgrades, new ways of looking at quality. Therefore, it is difficult to know the exact impact of effective procedures. Although not easy to calculate, the return on investment can be recognized by evaluating positive outcomes such as

- The catastrophic release that *didn't* occur
- A reduced number of lost-time accidents
- Less process downtime
- Higher productivity from a well-trained workforce
- Higher morale from workers who have been involved in writing and reviewing the procedures
- Less training time for new employees because the site has well-documented work activities (and a way to train on them)
- Better compliance with applicable regulations

- Smoother completion of infrequent tasks such as startup and shutdown

By tracking the role of effective procedures in continuous improvement of operations and maintenance, you can identify and recognize a positive return on your investment. In addition to beneficial financial results, effective procedures are a vital component for achieving and maintaining your process safety, quality, and environmental goals.

Endnotes

1. United States Department of Labor, Occupational Safety and Health Administration. *The Phillips 66 Company Houston Chemical Complex Explosion and Fire. A Report to the President.* Washington, DC, 1990.
2. United States Department of Labor, Occupational Safety and Health Administration. *Federal Register Volume 57 Number 36 pgs 6356-6417, 29 CFR Part 1910.119 Process Safety Management of Highly Hazardous Chemicals; Explosives and Blasting Agents, Final Rule.* Washington, DC, February 1992.
3. Keenan, Tom. "Process Safety Report Card." *Chilton's Industrial Safety and Hygiene News*. September 1994, pp. 19-20.
4. American Insurance Association. *Hazard Survey of the Chemical and Allied Industries*. American Insurance Association, 1979.
5. Marsh and McLennan Protection Consultants. *A Thirty Year Review of One Hundred of the Largest Property Damage Losses in the Hydrocarbon—Chemical Industries*. Marsh and McLennan, 1987.
6. American Institute of Chemical Engineers. *Software Directory*. American Institute of Chemical Engineers, New York, New York, 1995.

APPENDIX A

SELECTED PROCEDURE INITIATIVES, CONSENSUS CODES, AND REGULATIONS AFFECTING PROCEDURES

Organization	Selected Documents or Program Initiatives
AIChE Center for Chemical Process Safety (AIChE/CCPS) 345 East 47th Street New York, New York 10017.	*Plant Guidelines for Technical Management of Chemical Process Safety:* New York: AIChE/CCPS, 1992. *Guidelines for Technical Management of Chemical Process Safety* New York: AIChE/CCPS, 1989.
American Petroleum Institute (API) 1220 L Street NW Washington, DC 20005.	*API Recommended Practice 750: Management of Process Hazards.* Washington, DC: American Petroleum Institute, 1990.
Chemical Manufacturers Association (CMA) 2501 M Street, NW Washington, DC 20037.	*Responsible Care®: A Public Commitment.* Washington, DC: Chemical Manufacturers Association, 1988. *Responsible Care®: Process Safety Code of Management Practices.* Washington, DC: Chemical Manufacturers Association,1992.
Federal Emergency Management Agency (FEMA) 500 C Street SW Washington, DC 20472.	*Handbook of Chemical Hazard Analysis Procedures:* Washington, DC: FEMA, October 1990.

Organization	Selected Documents or Program Initiatives
Institution of Chemical Engineers, UK (IChemE) 165–171 Railway Terrace Rugby, Warks CV 21 3HQ, UK (0788) 78124.	*Loss Prevention Bulletin* (Case Histories) Conference on Major Accidents Prevention
International Labor Office (ILO) Geneva, Switzerland	*ILO Code of Practice for the Prevention of Major Industrial Accidents:* Geneva: International Labor Office, 1991.
International Organization for Standardization (ISO) Geneva, Switzerland or in United States of America American National Standards Institute (ANSI) 9000-1, 9001, 9002, 9003, 9004-1.	*Quality Management and Quality Assurance Standards* (9000) ISO 9000—ISO 9004 *Quality Systems—Models for Quality Assurance* (9001-9003) *Quality Management and Quality System Elements—Guidelines* (9004) Geneva, Switzerland, 1987.
Major Industrial Accident Council of Canada 600-265 Carling Avenue Ottowa, Ontario, Canada K1S 2E1.	Adopted AIChE/CCPS as voluntary code.
Official Journal of the European Communities	*Council Directive of 24 June 1982 (the Seveso Directive) on the Major-Accident Hazards of Certain Industrial Activities*. Berlin: Official Journal of the European Communities, 1989.
Organization for Economic Cooperation and Development (OECD)	*Workshop on Prevention of Accidents Involving Hazardous Substances Good Management Practice*, Discussion Document Relating to the Development of OECD Guideline for the Prevention of Accidents Involving Hazardous Substances. Berlin: Organization for Economic Cooperation and Development, 1989.
Organization Resources Counselors, Inc. (ORC) 1910 Sunderland Place NW Washington, DC 20036	*Recommendations for Process Hazards Management of Substances with Catastrophic Potential*. OSHA Docket S-026—Process Safety Management of Highly Hazardous Chemicals. Washington, DC 1988.

Organization	Selected Documents or Program Initiatives
The World Bank International Bank for Reconstruction and Development 1818 H Street, NW Washington, DC 20433	*Manual of Industrial Hazard Assessment Techniques*. Washington DC: The World Bank, 1985. Workshop Procedures Safety Management and Risk Control Occupational Health and Safety Guidelines
US Department of Labor (DOL) Occupational Health & Safety Administration 200 Constitution Avenue, NW Washington, DC 20210	*The Phillips 66 Company Houston Chemical Complex Explosion and Fire: A Report to the President*. Washington, DC: Occupational Health and Safety Administration, April 26, 1990. *Federal Register*, Volume 57 Number 36, pp. 6356–6417, "29 CFR 1910.119: Process Safety Management of Highly Hazardous Chemicals; Explosives and Blasting Agents." Final Rule. Washington, DC: February 1992.
US Environmental Protection Agency (EPA) OS-120 401 M Street SW Washington, DC 20460	*Federal Register*, Volume 58, Number 201, pp. 54190–54219. "Proposed Rule 40 CFR part 68: Risk Management Programs for Chemical Accidental Release Prevention." Washington, DC, October 1993.

APPENDIX B

COMMON POINTS OF API, OSHA, AND EPA

This appendix shows how the major elements of process safety management are reflected in API's RP 750, OSHA PSM Rule, and Environmental Protection Agency (EPA) regulations. The common points of each of the three documents are listed.

API RP 750 Section 5 *American Petroleum Institute Recommended Practice 750*	**OSHA 29 CFR 1910.119** *Process Safety Management of Highly Hazardous Chemicals, Explosives, and Blasting Agents*	**EPA 40 CFR Part 68.28** *Risk Management Programs for Chemical Accidental Release Prevention: Proposed Rule*
5.1. Content of Operating Procedures a. Identify person(s) responsible for each operations area		Document the safe and proper way to operate and maintain processes and equipment
b. Clear instructions consistent with process safety information	(f)(1) Develop and implement written operating procedures that provide clear instructions for safely conducting activities involved in each covered process consistent with the process safety information.	(b) Develop and implement written operating procedures that provide clear instructions for safely conducting activities involved in each covered process consistent with the process safety information.
c. Operating conditions and steps for following phases of operation: 1. Initial startup 2. Normal operations 3. Temporary operations 4. Emergency operations including: Emergency shutdowns and initiating persons 5. Normal shutdown 6. Startup following turnaround	(i) Steps for each operating phase: (A) Initial startup (B) Normal operations (C) Temporary operations (D) Emergency shutdown initiating conditions and assignment responsibilities (E) Emergency operations (F) Normal shutdown (G) Startup following turnaround or emergency shutdown	(1) Steps for each operating phase: (i) Initial startup (ii) Normal operations (iii) Temporary operations (iv) Emergency shutdown initiating conditions and assignment responsibilities (v) Emergency operations (vi) Normal shutdown (vii) Startup following turnaround or emergency shutdown
d. Operating limits 1. Consequences of deviation 2. Steps to correct or avoid deviation 3. Safety systems and their functions	(ii) Operating limits (A) Consequences of deviation (B) Steps to correct deviation	(2) Operating limits (i) Consequences of deviation (ii) Steps to correct deviation
e. Occupational safety and health considerations: 1. Properties and hazards 2. Precautions to prevent exposure, including engineering controls and PPE 3. Measures in event of contact or exposure 4. Special or unique hazards	(iii) Safety and health considerations: (A) Properties and hazards (B) Precautions to prevent exposure including engineering controls, administrative controls, and personal protective equipment (PPE) (C) Control measures in event of contact or airborne exposure (D) Quality control for raw materials and control of hazardous chemical inventory levels (E) Any special or unique hazards	(3) Safety and health considerations: (i) Properties and hazards (ii) Precautions to prevent exposure including engineering controls, administrative controls, and personal protective equipment (PPE) (iii) Control measures in event of contact or airborne exposure (iv) Quality control for raw materials and c ontrol of regulated substance inventory levels (v) Any special or unique hazards

d. Operating limits 3. Safety systems and their functions	(iv) Safety systems and their functions	(4) Safety systems and their functions
5.2. Operating Procedures in place before startup	(2) Procedures readily accessible	(c) Procedures readily accessible
5.3. • Reviewed in event of technology or facility changes. • Periodically reviewed to ensure current and accurate (3 to 5 years) • Review frequency based on degree of hazard presented	(3) • Procedures reviewed as necessary to ensure they reflect current operating practices • Certify annually that operating procedures are current and accurate	(d) • Procedures reviewed as necessary to ensure they reflect current operating practices • Certify annually that operating procedures are current and accurate
	(4) Develop and implement safe work practices: —lockout —confined space entry —opening process equipment or piping —entrance control Safe work practices shall apply to employees and contract employees	(e) Develop and implement safe work practices: —lockout —confined space entry —opening process equipment or piping —entrance control Safe work practices shall apply to employees and contract employees
8.4. Maintenance Systems Maintenance systems should include: • Maintenance procedures and safe work practices that ensure the mechanical integrity of equipment 8.5. Testing and Inspection program shall include: b. Testing and inspection procedures d. Procedures to correct equipment deficiencies or operations that are outside acceptable limits	(j) Mechanical integrity. (2) Establish and implement written procedures to maintain the on-going integrity of process equipment. (4) Inspection and testing. (ii) Inspection and testing procedures shall follow recognized and generally accepted good engineering practices.	68.32. Prevention Program—Maintenance (Mechanical integrity) (c) Establish and implement written procedures to maintain the on-going integrity of process equipment. (e) Maintenance Inspections (2) Inspection and testing procedures shall follow recognized and generally accepted good engineering practices

APPENDIX C

HOW TO DETERMINE THE TASKS THAT REQUIRE WRITTEN PROCEDURES

Generally, written operating and maintenance procedures are required by sound business practices or government mandate. Effective procedures provide written, clear instruction to the employee concerning how a task is safely and correctly performed. This helps to ensure that a task is performed with consistency regardless of the experience level of the employee.

There are different resources for identifying and selecting tasks that will require written procedures. Training Needs Assessments, Process Hazards Analyses, and Job Safety Analyses can assist you in analyzing the specific requirements of your plant.

Assessing Training Needs

Part of assessing training needs is determining the tasks to be performed for a given job position or area. Task Lists can be used to identify which procedures you need to develop. To generate a Task Listing, you identify the tasks performed for a specific job position or job area. A task is defined as a meaningful unit of work with a measurable value. A task contains more than two steps with a clearly defined starting and stopping point. For example, "mowing the lawn" is a task. The steps required to perform the task of "mowing the lawn" in a safe and prescribed manner begin to provide the detail for a written procedure.

You write procedures for safety, environmental, quality, and regulatory reasons. Regardless of the reason, you can determine the tasks (and the steps used to complete those tasks) for a specific job position or job area by:

1. Reviewing information related to the job area or position, including but not limited to, existing procedures, training, job safety analyses, process safety information, plant history, related industrial experience, and plant simulations, and incident reports.
2. Observing the job being performed by a competent person; recording the tasks and steps as they are actually performed.
3. Comparing your task listing to the existing documents to verify that you have not missed any tasks.

When you have completed documenting your task listing, most of tasks listed will become procedures titles. The task listing must be evaluated based on specific criteria including criticality, complexity, frequency of performance and potential hazard. The evaluation determines which tasks become written procedures. Although the task steps are not usually detailed enough to be procedures, they provide the foundation for procedure development.

Process Hazards Analyses

You often identify critical, difficult, or hazardous tasks as part of a Process Hazards Analysis (PHA). PHA information is another resource used to indicate which procedures need to be written. The PHA also directs you to tasks that may require much more detailed procedures or training because of criticality or safety issues.

Job Safety Analyses

Job Safety Analyses identify safety and industrial hygiene issues for a particular job position, area, or tasks. They are a resource for determining which tasks should be described by written procedures. Job Safety Analyses can also identify special cautions and warnings that should be included in a procedure.

APPENDIX D

PROCEDURE PERFORMANCE EVALUATION

This appendix includes a sample procedure performance evaluation document which can be used by an evaluator to evaluate a trainee's performance of a procedure. This document, when used in conjunction with an approved, up to date procedure, becomes a performance evaluation tool. The items listed on the second page should be customized to your facility's training program requirements.

How will you verify that a trainee can perform a task (as written in a procedure) safely and correctly? You can use the procedure performance evaluation with the procedure while observing the trainee perform or simulate the procedure. The procedure is the criteria for safe and correct performance of the task, and the procedure performance evaluation document is the guideline for the evaluator and the trainee. Upon the completion of the performance evaluation, this document becomes the record of performance.

PROCEDURE PERFORMANCE EVALUATION[1]

Objective:

- The Trainee will demonstrate the ability (actual or simulated) to safely and correctly perform the operations required by the procedure, using the procedure and Qualification Card.

[1]Reprinted by permission of NUS Training Corporation, Gaithersburg, MD, 1992.

Instructions to Observer:

- Review procedure and checklist with the trainee.
- Observe trainee performing (or simulating walkthrough/oral) procedure steps (trainee may use a working copy of the procedure.)
- When you observe correct performance of a step or substep, check and initial the step or substep.
- If an action is omitted or performed incorrectly, circle that action on the procedure. Discuss the omission or error with the trainee before continuing the qualification session.
- If the trainee attempts an unsafe action or ignores a potentially hazardous condition, STOP the trainee and terminate the qualification session. Discuss the violation with the trainee before allowing the trainee to reschedule another qualification.
- If specifications are required (for instance: tanks, quantities, valve line-ups), give the trainee appropriate instructions so the session can continue.
- If the procedure requires permission or a request, grant the request or give permission.
- Do not prompt the trainee or allow anyone to assist the trainee unless required by the procedure or a condition exists which would normally require assistance.
- If a procedure step requires the trainee to verify a condition but you cannot directly observe whether the trainee has checked it, verify the condition yourself and check the step in the procedure if the condition is actually satisfied.
- When the procedure is completed, discuss any omissions or problems encountered during the performance (simulation).
- Complete the qualification card.
- If the trainee satisfactorily performed (simulated) the procedure, both you and trainee sign the qualification card.
- Attach the procedure to Procedure Performance Evaluation and return to Training Department.

Instruction to Trainee:

"I am going to observe you ______________________. I will follow your performance using the procedure. While performing the procedure, you may use any references you want, including a copy of the procedure for ______________. You must perform the task alone, with no help from anyone else, unless assistance is specified in the procedure or there is an abnormal or emergency condition. Do you have any questions?"

DATE QUALIFICATION STARTED:______________ DATE COMPLETED: ______________

OBSERVER'S NAME (print) ________________________ EMP. NO.:______________

TRAINEE'S NAME (print)____

PROCEDURE TITLE, NO., & REVISION NO.: ______________________________________

Did the trainee:	YES	NO	NA	COMMENTS include number of omitted or incorrectly performed steps
1. Obtain all special tools or equipment required by the procedure?				
2. Verify all prerequisites satisfied?				
3. Follow procedure steps in the order written?				
4. Obtain required permissions?				
5. Obtain/wear all required personal protective equipment?				
6. Follow all safety requirements?				
7. Verify all valve lineups?				
8. Know the location of: • Local start/stop switches? • Valves? • Controllers? • Control boards and panels? • Instrumentation? • Equipment? • Emergency stops? • Remote start/stop switches? • Safety equipment				
9. Demonstrate ability to safely and properly use or operate auxiliary equipment (cranes, hoists,etc.)?				
10. Know actions to take in event of alarms or malfunctions?				

______________________________ successfully demonstrated the ability to safely and correctly perform this procedure.

______________________________ DID NOT successfully demonstrate the ability to safely and correctly perform this procedure because (clearly state reason):

TRAINEE'S SIGNATURE ________________________________ DATE ____________

OBSERVER SIGNATURE________________________________ DATE ____________

ATTACH copy of procedure used to verify performance to Procedure Performance Evaluation and RETURN documents to Training Dept.

APPENDIX E

PROCEDURE CRITERIA CHECKLIST

This appendix lists some Procedure Evaluation Criteria which are based on procedure writing experience, human factors, and lessons learned including recognized formatting and presentation methods that will help to ensure your procedures are effectively written. While most procedure elements are based on human factors, some are based on technical, administrative, or regulatory factors. Revise this list to reflect your facility's requirements and include it in your procedure management system.

PROCEDURE CRITERIA CHECKLIST[1]

PROCEDURE CRITERIA CHECKLIST ITEMS	YES	NO/NA
1. Does the title accurately describe the nature of the activity?		
2. Is the necessary procedure control information included on each page such as? • Facility or unit name or identifier • Procedure title • Procedure number • Date of issue, approval date, required review date, and effective date • Revision number • Page number and total pages		
3. Is the last page of the procedure clearly identified?		
4. Are temporary procedures clearly identified?		
5. Does every procedure have a unique and permanent identifier?		

PROCEDURE CRITERIA CHECKLIST ITEMS	YES	NO/NA
6. Does the procedure Purpose clearly describe the intent or goal of the procedure?		
7. For duplicate processes, are the procedures complete and accurate for each process?		
8. Is all information necessary for performing the procedure included or referenced in the procedure?		
9. Are all items referenced in the procedure listed in the References section of the procedure?		
10. Are items listed in the References section of the procedure correctly and completely identified?		
11. Are procedures written in upper- and lowercase letters?		
12. Are Cautions, Warnings, and Notes placed immediately before the step to which they apply?		
13. Do Cautions, Warnings, and Notes stand out from procedure steps?		
14. If more than one person is required to perform the procedure, is the person responsible for performing each step identified?		
15. Is a signoff line provided for verifying critical steps of a procedure? (Optional)		
16. If the procedure requires coordination with others, does it contain a checklist, signoff, or other method for indicating the steps or actions have been performed or completed?		
17. Are steps written in short, concise sentences?		
18. If a step contains more than two items, are they listed rather than buried in the text?		
19. If two actions are included in a single step, can the actions actually be performed simultaneously or as a single action?		
20. Are steps that must be performed in a fixed sequence identified as such?		
21. Are operating or maintenance limits or specifications written in quantitative terms?		
22. Does the procedure provide instructions for all reasonable contingencies? If contingency instructions are used, does the contingency statement precede the action statement?		

PROCEDURE CRITERIA CHECKLIST ITEMS	YES	NO/NA
23. If conditions or criteria are used to help the user make a decision or recognize a situation, do they precede the action?		
24. Are conditional instructions easy to understand? If an action must meet more than two requirements, are the requirements listed?		
25. Are calculations clear and understandable? For complicated or critical calculations, is a formula or table included or referenced?		
26. Can graphs, charts, and tables be easily and accurately extracted and interpreted?		
27. Do procedures that specify alignment such as valve positions, pipe and spool configurations, or hose station hook-ups • Specify each item • Identify each item with a unique number or designator • Specify the position in which the item is to be placed • Indicate where the user records the position if applicable		
28. Do Emergency Operating Procedures contain provisions for verifying • Conditions associated with an emergency (initiating conditions) • Automatic actions associated with an emergency • Performance of critical actions		
29. Do Maintenance Procedures include required followup actions or tests and tell the user who must be notified?		
30. If a procedure must be performed by someone with a special qualification, are the required technical skill levels identified?		

Endnote

1. Adapted from the NUS Training Corporation's *Procedure Writing Workshop Manual*, Seventh Edition, Gaithersburg, MD, 1995.

APPENDIX F

SAMPLE PROCEDURE FORMATS

The sample procedure formats in this appendix are derived from actual procedures used in industry. They show various ways of presenting information in a T-format. Although each format is slightly different, they all include:

- Identifying information
- Unique title
- Introductory sections
- Page numbers
- Revision numbers
- Approval
- Date

ANY PLANT **PLANT LOCATION** **SOP—001 REV. NO. 3**

procedure identifier/revision number

STARTING XYZ AFTER COLD SHUTDOWN

PURPOSE: Xxxxx xxx x xxxx xxxx. X XXXX xxx xxx xxxxx xxxx. Xxxx xxx xxx xx x xxx xx x xx x xxx. Xxxx xxx x x xxx x xx111 xx xxxxx.

REFERENCES: MSDS (Material Safety Data Sheets)

Chemical name	*Name in procedure*
Vinyl Acetate	[MVAC]
Sulfuric Acid Fuming	[Oleum]
Petroleum Jelly	[Jam]

Signatures
Production Superintendent
Safety Engineer
Safety Supervisor
Training Supervisor
Production/Technical Manager
Plant Manager Approval

PRECAUTIONS: Xxxxx xxx x xxxx xxxx. X XXXX xxx xxx xxxxx xxxx. Xxxx xxx xxx xx x xxx xx x xx x xxx. Xxxx xxx x x xxx x xx 111 xx xxxxx.

ABBREVIATIONS:

BMT Batch Mix Tank
Xxx XX xxxx xxx xxxx

SAFETY SYSTEMS/INTERLOCKS:

Interlock 1.	set point
High level alarm	set point
DBS Snuffer System	initiator set point

PREREQUISITES: Xxxxx xxx x xxxx xxxx. X XXXX xxx xxx xxxxx xxxx. Xxxx xxx xxx xx x xxx xx x xx x xxx. Xxxx xxx x x xxx x xx111 xx xxxxx.

PROCEDURE STEPS: NOTE: X xxx xxxx xxx. X xx xxx xxx x xxx xx x. Xx xxx xxx.

1. XXXX xxxx xxxx.
2. XXXX xxx xxx xx.

END

ANY PLANT PLANT LOCATION OPS-SUP-001 REV. 3

PROCEDURE TITLE

PURPOSE: Xxxxx xxx x xxxx xxxx. X XXXX xxx xxx xxxxx xxxx. Xxxx xxx xxx xx x xxx xx x xx x xxx. Xxxx xxx x x xxx x xx111 xx.

PRECAUTIONS: Xxxxx xxx x xxxx xxxx. X XXXX xxx xxx xxxxx xxxx. Xxxx xxx xxx xx x xxx xx x xx x xxx. Xxxx xxx x x xxx x xx111 xx.

REFERENCES: Xxxxx xxx x xxxx xxxx. X XXXX xxx xxx xxxxx xxxx. Xxxx xxx xxx xx x xxx xx x xx x xxx. Xxxx xxx x x xxx x xx111 xx.

SPECIAL EQUIPMENT: Xxxxx xxx x xxxx xxxx. X XXXX xxx xxx xxxxx xxxx. Xxxx xxx xxx xx x xxx xx x xx x xxx. Xxxx xxx x x xxx x xx111 xx.

PREREQUISITES: Xxxxx xxx x xxxx xxxx. X XXXX xxx xxx xxxxx xxxx. Xxxx xxx xxx xx x xxx xx x xx x xxx. Xxxx xxx x x xxx x xx111 xx.

PROCEDURE STEPS: NOTE: X xxx xxxx xxx. X xx xxx xxx x xxx xx x. Xx xxx xxx xx xxx xxxx.

1. XXXX xxxx xxxx.

CAUTION

X xxx xxxx xxx. X xx xxx xxx x xxx xx x. Xx xxx xxx xx xxx xxxx.

CAUTION

2. XXXX xxx xxx xx.
3. XXXXX xx xxx:
   a. XXX1
   b. XXY1
4. XXXX xx xxxxx.
5. XXXX x xxx:
   —123X
   —124X
   —125X

o O o

Date Issued:____________ OPS-SUP-001 Rev. 2. Page 1 of 1

Required Review Date:____________

Unit Manager [approval appears on first page only]______________________________

COMPANY NAME
STANDARD OPERATING PROCEDURES
PLANT NAME

CCR UNIT NO. 2. CCR2-SOP-37 Revision 2

TITLE: Transferring XYZ from Storage Tanks To Blenders

AUTHORIZED BY: *Signature/Date Reviewed*

Safety Manager ____________________

Training Supervisor ____________________

Environmental Manager ____________________

Production Manager—Unit (approval) ____________________

Engineering Manager ____________________

Plant Manager ____________________

PURPOSE: Xxxxx xxx x xxxx xxxx. X XXXX xxx xxx xxxx xxxx. Xxxx xxx xxx xx x xxx xx x xx x xxx. Xxxx xxx x x xxx x xx111 xx xx..

REFERENCES: Xxxxx xxx x xxxx xxxx. X XXXX xxx xxx xxxx xxxx. Xxxx xxx xxx xx x xxx xx x xx x xxx. Xxxx xxx x x xxx x xx111 xx xx.

SAFETY EQUIPMENT: Xxxxx xxx x xxxx xxxx. X XXXX xxx xxx xxxx xxxx. Xxxx xxx xxx xx x xxx xx x xx x xxx. Xxxx xxx x x xxx x xx111 xx xx.

HEALTH HAZARDS: Xxxxx xxx x xxxx xxxx. X XXXX xxx xxx xxxx xxxx. Xxxx xxx xxx xx x xxx xx x xx x xxx. Xxxx xxx x x xxx x xx111 xx xx.

PROTECTIVE SAFETY EQUIPMENT: Xxxxx xxx x xxxx xxxx. X XXXX xxx xxx xxxx xxxx. Xxxx xxx xxx xx x xxx xx x xx x xxx. Xxxx xxx x x xxx x xx111 xx xx.

1.0. PRE-START-UP PROCEDURE STEPS:
   - 1.1. XXXX xx xxxx.
      - 1.1.1. XXX xxx x.
      - 1.1.2. XXXXX xxx xxxx.
      - 1.1.3. XXX xxx x.
   - 1.2. XXX xx xxx.
   - 1.3. XXXX xxx xxxx.

2.0. ROUTINE PROCEDURE STEPS:
   - 2.1. XXXX xx xxxx.
   - 2.2. XXX xx xxx.

PROCEDURE COMPLETE

Date of Issue Required Review Date Page 1 of 1

ANY PLANT **PLANT LOCATION** **SOP-223 OPS REV. 1**

Approved for Use: DATE EFFECTIVE________________

Unit Manager [approval appears on first page only]

PROCEDURE TITLE

PURPOSE: Xxxxx xxx x xxxx xxxx. X XXXX xxx xxx xxxxx xxxx. Xxxx xxx xxx xx x xxx xx x xx x xxx. Xxxx xxx x x xxx x xx111 xx.

PRECAUTIONS: Xxxxx xxx x xxxx xxxx. X XXXX xxx xxx xxxxx xxxx. Xxxx xxx xxx xx x xxx xx x xx x xxx. Xxxx xxx x x xxx x xx111 xx.

REFERENCES: Xxxxx xxx x xxxx xxxx. X XXXX xxx xxx xxxxx xxxx. Xxxx xxx xxx xx x xxx xx x xx x xxx. Xxxx xxx x x xxx x xx111 xx.

SPECIAL EQUIPMENT: Xxxxx xxx x xxxx xxxx. X XXXX xxx xxx xxxxx xxxx. Xxxx xxx xxx xx x xxx xx x xx x xxx. Xxxx xxx x x xxx x xx111 xx.

ALARMS/SETPOINTS:

ALARM	*SET POINT*
AAA Hi	120 units
AAA Lo	80 units
BBB Hi-Hi	90 units interlocks CCC

PREREQUISITES: Xxxxx xxx x xxxx xxxx. X XXXX xxx xxx xxxxx xxxx. Xxxx xxx xxx xx x xxx xx x xx x xxx. Xxxx xxx x x xxx x xx111 xx.

PROCEDURE STEPS: NOTE: X xxx xxxx xxx. X xx xxx xxx x xxx xx x. Xx xxx xxx xx xxx xxxx.

1. XXXX xxxx xxxx.

CAUTION

X xxx xxxx xxx. X xx xxx xxx x xxx xx x. Xx xxx xxx xx xxx xxxx.

CAUTION

2. XXXX xxx xxx xx.
3. XXXXX xx xxx:

o O o

COMPANY NAME **(Issue Date:)**

Title:	SHUT DOWN High Pressure Storage Tank	NT-2.2
Area:	Nitric Terminal	Page 1 of 1

[In this example, procedure approval signatures appear on the first sheet of the operating manual.]

1.0. GENERAL DESCRIPTION

Xxxxx xxx x xxxx xxxx. X XXXX xxx xxx xxxxx xxxx. Xxxx xxx xxx xx x xxx xx x xx x xxx. Xxxx xxx x x xxx x xx111 xx xxxxx.

2.0. USER

Operator

Maintenance Mechanic

3.0. REFERENCES

Safety Manual

Material Safety Data Sheets

4.0 SAFETY

Xxxxx xxx x xxxx xxxx. X XXXX xxx xxx xxxxx xxxx. Xxxx xxx xxx xx x xxx xx x xx x xxx. Xxxx xxx x x xxx x xx111 xx xxxxx.

5.0 REQUIRED EQUIPMENT

Xxxxx xxx

X xxxx xxxx

6.0 PREREQUISITES

None Applicable

7.0 INSTRUCTIONS

SAFETY NOTE:	X xxx xx x xxx. Xxx xxx x xxxx. Xxx xxxx xxx.

7.1. Xxxx-xxx xx xxx 123.

7.1.1. Xxxx xxx xx xxx.

7.7.2. X xx xxxx x.

7.2. XXX x xxx x.

7.2.1. XXX x xx xx.

END

APPENDIX G

SAMPLE FORMATS OF OPERATING LIMITS TABLES

The following appendix includes two samples for formatting operating limits tables. These can be customized to meet your facility's specific requirements. These samples also demonstrate the interaction between operating limits tables and procedures. The required actions in the tables can either be the actual required steps with all necessary explanatory information or a reference to the applicable procedure. The steps or references to procedures should follow the development criteria listed in Chapter 5.

Process B Operating Limits Table

Process Variable (variable and monitoring location)	Normal Operating Range (units)	Alarms	Consequences of Deviating Outside of Normal Operating Range	Required Actions	Safe Upper Limit	Safe Lower Limit	Consequences of Deviating Outside of Safe Operating Limits	Required Actions
								Listed actions or reference to a procedure.

Tag/Loop # ________
System ________
Ref. P&ID# ________
PM# ________

STANDARD OPERATING CONDITIONS

ID# ________
Document # ________
Page 1 of ________
Equipment Name ________

Name/Location Description	Never Exceed		Interlocks		Alarms		Operating Range		Remarks: R—Reason for Limit or Effect of Exceeding Limit; C—Correction and/or Control Action; T—Troubleshooting Guide
	Min	Max	Min	Max	Min	Max	Min	Max	
	Where "NEVER EXCEED" limits are applicable, any deviation outside these limits will require a document incident investigation.								

Effective Date ________
Rev. No ________
Original Date ________

____
Area Supervisor
Next Review Date ________

GENERAL REFERENCES

American Insurance Association. *Hazard Survey of the Chemical and Allied Industries*. American Insurance Association, 1979.

American Petroleum Institute. *Management of Process Hazards American Petroleum Institute, API Recommended Practice 750*. Washington, DC, 1990.

Center for Chemical Process Safety. *Guidelines for Technical Management of Chemical Process Safety*. AIChE, 1989.

Center for Chemical Process Safety. *Plant Guidelines for Technical Management of Chemical Process Safety*. AIChE, 1992.

Chemical Manufacturers Association. *Process Safety Codes of Management Practices*. Washington, DC, 1990.

International Labor Office (ILO). *ILO Code of Practice for the Prevention of Major Industrial Accidents*. Geneva, Switzerland, 1991.

Keenan, T. "Process Safety Report Card." *Chilton's Industrial Safety and Hygiene News*. September, 1994, pp. 19-20.

Marsh and McLennan Protection Consultants. *A Thirty Year Review of One Hundred of the Largest Property Damage Losses in the Hydrocarbon—Chemical Industries*. Marsh and McLennan, 1987.

NUS Training Corporation. *Procedure Writing Workshop Manual*. Gaithersburg, MD, Seventh Edition, 1995.

Organization for Economic Cooperation and Development (OECD). *Discussion Document Relating to the Development of OECD Guidance for the Prevention of Accidents Involving Hazardous Substances* from the Workshop on Prevention of Accidents Involving Hazardous Substances Good Management Practices. Berlin, Germany, 1989.

United States Department of Labor, Occupational Safety and Health Administration. *The Phillips 66 Company Houston Chemical Complex Explosion and Fire. A Report to the President*. Washington, DC, 1990.

United States Environmental Protection Agency. *Federal Register* Volume 58 Number 201 pp. 54190–54219. "40 CFR Part 68: Proposed Risk Management Programs for Chemical Accidental Release Prevention." Washington, DC, 1993.

United States Department of Labor, Occupational Safety and Health Administration. *Federal Register* Volume 57 Number 36 pp. 6356–6417. "29 CFR Part 1910.119: Process Safety Management of Highly Hazardous Chemicals; Explosives and Blasting Agents, Final Rule." Washington, DC, February 1992.

INDEX

CPSIA information can be obtained at www.ICGtesting.com
Printed in the USA
BVOW07*0401201113

336688BV00008B/286/P